Understanding Geology Through Maps

Understanding Geology Through Maps

Graham Borradaile
Emeritus Professor
Lakehead University
Thunder Bay
Canada

AMSTERDAM • BOSTON • HEIDELBERG • LONDON • NEW YORK • OXFORD • PARIS
SAN DIEGO • SAN FRANCISCO • SINGAPORE • SYDNEY • TOKYO

ELSEVIER

Elsevier
Radarweg 29, PO Box 211, 1000 AE Amsterdam, Netherlands
The Boulevard, Langford Lane, Kidlington, Oxford OX5 1GB, UK
225 Wyman Street, Waltham, MA 02451, USA

Notices
Knowledge and best practice in this field are constantly changing. As new research and experience broaden our
understanding, changes in research methods, professional practices, or medical treatment may become necessary.

Practitioners and researchers must always rely on their own experience and knowledge in evaluating and using any
information, methods, compounds, or experiments described herein. In using such information or methods they
should be mindful of their own safety and the safety of others, including parties for whom they have a professional
responsibility.

To the fullest extent of the law, neither the Publisher nor the authors, contributors, or editors, assume any liability
for any injury and/or damage to persons or property as a matter of products liability, negligence or otherwise, or
from any use or operation of any methods, products, instructions, or ideas contained in the material herein.

ISBN: 978-0-12-800866-9

Library of Congress Cataloging-in-Publication Data
Borradaile, G. J.
    Understanding geology through maps / Graham Borradaile. -- First edition.
        pages cm
    ISBN 978-0-12-800866-9
1. Geology--Maps. I. Title.
    QE36.B67 2014
    551--dc23
                        2014006592

British Library Cataloguing in Publication Data
A catalogue record for this book is available from the British Library

For information on all Elsevier publications
visit our web site at store.elsevier.com

This book has been manufactured using Print On Demand technology.

Working together
to grow libraries in
developing countries

www.elsevier.com • www.bookaid.org

Contents

The geological terminology used in a course using this material is minimal and much less difficult than the new terminology required in many other science courses. The prerequisite or accompanying a geological map interpretation course would be a lecture course in first-year physical geology. For those who have not previously followed a geology course, the essential principles involved are explained as we progress. Definitions and terminology are presented incrementally and may be recognized as words or phrases that are explained in the Appendix. The full meaning or usage of terms may only become apparent later after completion of an exercise. The first three chapters essentially duplicate material from a first-year geology course.

Geological map interpretation should not be confused with cartography (drawing maps), GIS (geographical information systems), topographic (land surface) map interpretation, geomorphological (land-form) interpretation or with the interpretation of aerial photographs or satellite images. We shall learn that a map showing the distribution of rock types and their structures may yield the geological history, relative ages of rocks and events, useful volumes, and locations of economically valuable materials.

First, and unlike all other forms of map interpretation, geological map interpretation permits, and is constantly concerned with, the relationships and relative positions of rock units that are below the ground surface. Few nongeologists are aware of this distinction. At the very least, the three-dimensional interpretation of relative positions will be valid to a depth equal to the topographic relief. The popular book by Simon Winchester (2001) explains the philosophy and history of the geological map its development of the concept by William Smith in the 1780s.

Second, geological map interpretation constantly places geological structures and rock units in relative chronological order. Such relative age determination depends on the form of special geological map relationships and is quite independent of any laboratory determination of absolute ages (geochronology using radioactive decay). Nor does it require any knowledge of the fossils that sedimentary rock units may contain. However, such information may be included into problems for interest.

In some cases, the information needed may be more fragmentary than that present on a map; data from boreholes (borehole logs or well logs) or isolated outcrops commonly suffices to create a geological map, from which the geological history of the region may be understood.

There are more exercises than necessary to develop a working knowledge of geological map interpretation and the instructor may therefore omit material according to the availability of time and the appropriateness for the course.

In an a sociological era in which technology and data processing has provided the cutting edge for geological science, some students (and even some geologists) mistakenly regard pencil and paper drafting and the mental and physical manipulation of map, well log, and section information as an anachronism. Unfortunately, nothing could be further from the truth. Whereas the material may seem less profound to a superficial observer than that, for example, in structural geology, geophysics, or geochemistry, there is in fact no viable alternative for developing an understanding of geological maps and "reading" them to visualize the three-dimensional configuration of the rock units and structures. A student's most powerful intellectual tool is humility; if you commence this course and every other geology course with the same rigorous, scholarly attitude as with a course in calculus, you will develop a very powerful level of comprehension.

Whereas this may seem old fashioned, remember that technology does not replace the groundwork of many aspects of science. Just as with learning mathematics or microscopic work, etc. one has to exercise the brain in traditional intellectual paths and a computer does not spare us any effort nor put knowledge in our heads. Computer map interpretation is a commonly used expression but is somewhat of a misnomer. It is true that (with enormous effort), map information may be digitized, including the geological observations and well log data. This may be fed into a powerful computer program that produces cross-sections and even three-dimensional images of the rock relationships. Unfortunately, no one algorithm can satisfy all possible geological scenarios for different areas and different maps and the effort of digitizing and feeding the data into the program takes much longer than traditional intellectual methods of map interpretation. Remember that the computer is only as good as the program. Programs are justified in industry because such software is customized to the geology of the region and usually specific to a small area (mine or oil field) and the rewards are enormous. As new information is acquired during exploration or exploitations, the commercial geologists may update the program and instantly obtain revised cross-sections and three-dimensional images. However, one cannot even begin to understand the output of those programs without a thorough comprehension of the material handled in this course in a traditional manner such as presented in this book.

MINIMUM MATERIALS REQUIRED FOR THIS COURSE ARE

1. 2H pencil, HB pencil, eraser, some colored pencil crayons (the use of ball pen, ink, or felt marker is unacceptable, since many constructions require incremental improvements and changes).
2. Transparent plastic rule, 12 inches/30 cm.
3. One transparent plastic drawing triangle (45° or 60/30° are equally good).
4. Some tracing paper (approximately 8″ × 11″).
5. Some graph paper for accurate construction of cross-sections (a sample is included in the Appendix).

OTHER LITERATURE

With the exception of the books by Lisle, Maltman, and Thomas, the examples listed here mostly deal with simplified extracts from geological maps, in the form of black-and-white line diagrams. Simpler than real published geological maps, they are actually the only means to grasp a genuine understanding of three-dimensional relationships and the sequence of geological events. This book comprises both synthetic maps and simplified extracts from real geological maps.

FURTHER READING

Aitken, M.J., 1994. *Science based dating in Archaeology.* London and New York, Longman, 274 pp.

Anderson, E.M., 1951. *The dynamics of faulting,* Edinburgh, Oliver and Boyd, 206 pp.

Bennison, G.M. and Mosely, K.A., 2003. *An introduction to geological structures and maps.* J. Wiley, New York, 160 pp.

Butler, B.C.M. and Bell, J.D., 1988. *Interpretation of geological maps.* Longman, UK, 236 pp.

Lisle, R.J., 1988. *Geological structures and maps: a practical guide.* Pergamon Press, Oxford, 150 pp.

Maltman, A. 1991. *An introduction to geological maps.* Open University Press, 208 pp.

Means, W.D., 1976. *Stress and Strain: basic concepts of continuum mechanics for geologists,* New York, Springer-Verlag, 338 pp.

Park, R.G., 1989. *Foundations of structural geology.* Blackie, 148 pp.

Platt, J.I. and Challinor, J., 1968. *Simple geological structures.* George, Allen & Unwin, London, 57 pp.

Price, N.J. and Cosgrove, J.W., 1990. *Analysis of geological structures.* Cambridge University Press, 502 pp.

Price, N.J., 1966. *Fault and joint development in brittle and semi-brittle rock.* Oxford, Pergamon Press, 176pp.

Ragan, D.M., 1985. *Structural geology: an introduction to geometric techniques.* 3rd Ed., Wiley, 393 pp.

Ramsay, J.G. & Huber, M. I. 1983. The techniques of modern structural geology. *Strain analysis.* Volume I. Academic Press, 307 pages

Ramsay, J.G. 1967. *Folding and Fracturing of Rocks,* New York, McGraw-Hill, 578 pp.

Ramsay, J.G. and Lisle, R.J., 2000. The techniques of Modern structural geology. *Applications of continuum mechanics in structural geology.* Vol 3. Academic Press, 304 pp.

Simpson, B., 1968. *Geological maps.* Pergamon Press, Oxford, 98 pp.

Thomas, W.A., 2004. *Meeting challenges with geological maps.* American Geological Institute. Alexandra, Virginia, USA, 65pp. ISBN: 0-922152-70-5.

Winchester, S., 2001. *The map that changed the World.* New York, Harper Collins, 329 pp.

ADVANCED STRUCTURAL INTERPRETATION OF GEOLOGICAL MAPS

These are mentioned for completeness and for the benefit of Geology and Earth Science majors who will study Structural Geology. They are listed in order of increasing difficulty.

Marshak, S. and Mitra, G., 1998. *Basic methods of Structural Geology.* Prentice Hall, New Jersey, 446 pp.

Ragan, D.M., 1985. *Structural Geology: An introduction to geometrical techniques.* J. Wiley, New York, 393 pp.

Ramsay, J.G. and Huber, M.I., 1987. *The techniques of modern structural geology.* Vol. 2, J. Wiley, New York, 391 pp.

Worldwide stratigraphic nomenclature and International Commission base ages				
Era	Period	≥Ma (2002) base	≥Ma (1985) base	Principal protolith for UK and Ireland geological map
Cenozoic	Neogene and Palaeogene	65	65	Marine clay and sand (glacial sediment omitted)
Mesozoic	Cretaceous	142	140	Chalk
	Jurassic	206	195	Limestone
	Triassic	248	230	"New red" sandstone
Paleozoic	Permian	290	280	Evaporite "new red" sandstone
	Carboniferous	350	345	Limestone, coal, gritstone
	Devonian	417	395	"Old Red" sandstone
	Silurian	443	445	Slate, greywacke limestone
	Ordovician	495	510	Slate
	Cambrian	545	570	Slate, sandstone, limestone
Pre-Cambrian	Proterozoic and Late Archean	>545	>570	Many, mostly metamorphosed

FIGURE 0.1 An example of a simplified stratigraphic table.

Acknowledgments

Numerous graduate assistants including Bjarne Almqvist, France Lagroix, Tomasz Werner, John Dehls, Bob Spark, John McArthur, David Gautier, Dawn-Anne Trebilcock, and many more helped me in the running of an elementary map interpretation course on which this book is based. Sam Spivak drew some of the excerpts from published geological maps, originally in color. I have simplified these and drawn them using "CorelDraw" in black and white for this text. Ornaments and patterns used to indicate rock types are from Pangaea Scientifics' Rockfill collection.

Excerpts from published geological maps are used with permission of the United States Geological Survey, Geological Survey of Canada, the British Geological Survey, the Geological Survey of Sweden, Geological Survey of Manitoba, Ontario Geological Survey, Geological Survey of Israel, and the Geological Survey Department of Cyprus.

Geological Maps and Some Basic Terminology

William Smith (1736–1839) created the first geological map in southern England (see Simon Winchester's book, 2001 listed in the Foreword). He expanded it through his lifetime's work to be a geological map of England and Wales which is little different from the map we know today (Figure 1.1 shows the modern geological map of the United Kingdom). His initial map focused on a small area around Bath (Figure 1.2). First appearing and acknowledged in 1799, and published in 1802, the original geological map provided a revolutionary breakthrough in that it provided a way of showing map distribution of rock types from which one could deduce their relative age (≈chronostratigraphic sequence) and three-dimensional configuration at depth from visual inspection of the surface map alone. The father of the geological map, Smith, grew up in a rural environment in Oxfordshire but was exposed to geology at an early age through an interest in the locally abundant fossils. Parenthetically, a wealthy nephew John Phillips became the first Professor of Paleontology and Geology at Oxford and in later life assisted his uncle from discrediting attacks by formally educated intellectuals. Most of the population denied the antiquity of fossils due to religious reasons or they attributed their presence to the effects of the biblical flood. However, William Smith, like other better educated and more fortunate natural philosophers, grew to suspect the great ages required for the accumulation of thick sequences of sediment (later to be lithified to become sedimentary rock). Despite the controversy about their origin,

Smith was the first person to recognize that specific fossils characterized each stratigraphic horizon. A further quantum step in knowledge was made when he realized that the fossil content could then be used to correlate comparable sequences of rocks between different locations (Figure 1.3). The details for central and southern England differ little from Smith's final compilation, which was published in 1815. With the benefit of recent geochronological work, the absolute ages of the geological periods have been established (Figure 1.4).

Smith introduced the term stratigraphical and stratigraphy about 1795. The worldwide stratigraphic column as we now know it is shown in Figure 1.4. One of the few pieces of memory work in studying this book is to become familiar with the sequence of names and their approximate geochronological ages. Stratum (plural strata) derives from the Latin for street, since the well-known Roman Roads of Britain (and elsewhere in Europe) have survived two millennia due to their well designed layered structure, from the foundation of gravel, through a sand layer, to cobbles at the top. Bed and bedding are more-or-less English synonyms for stratum and stratification, although strictly speaking "bed" would describe a sedimentary rock, whereas stratification also encompasses sequentially layered igneous rocks such as lavas and even layered magmatic rock.

Bedding plane is subtly and importantly different. It refers to the discrete horizon of no thickness, a stratum-parallel boundary that separates two different beds,

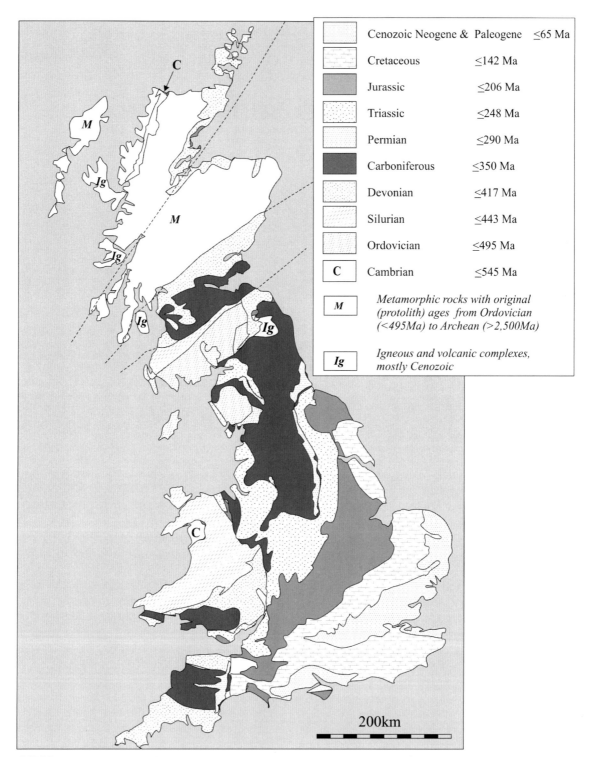

Cenozoic Neogene & Paleogene   ≤65 Ma
Cretaceous   ≤142 Ma
Jurassic   ≤206 Ma
Triassic   ≤248 Ma
Permian   ≤290 Ma
Carboniferous   ≤350 Ma
Devonian   ≤417 Ma
Silurian   ≤443 Ma
Ordovician   ≤495 Ma
C   Cambrian   ≤545 Ma

M   *Metamorphic rocks with original (protolith) ages from Ordovician (<495Ma) to Archean (>2,500Ma)*

Ig   *Igneous and volcanic complexes, mostly Cenozoic*

200km

FIGURE 1.1   Geological map of Scotland, England, and Wales. England and Wales were the first countries to be mapped in this way.

e.g., between a bed of sandstone and a bed of limestone. A bedding plane represents some unknown interval of nondeposition that may have endured for seconds or thousands of years. James Hutton (1726–1797) recognized the importance of a very special class of bedding surface, the

unconformity. Hutton, quite unlike Smith, was a formally educated scholar and a medical doctor and a Professor at the University of Edinburgh at the acme the intellectual interval known as the period of Scottish Enlightenment. Hutton observed tilted strata, eroded, and subsequently

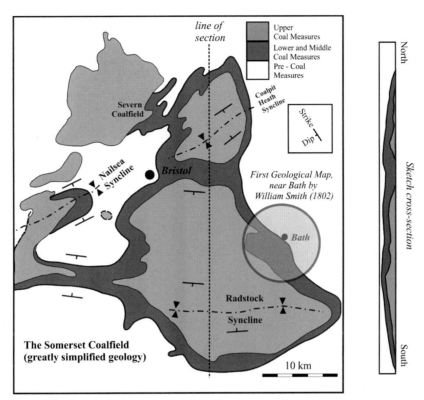

FIGURE 1.2   The small circular area around Bath was the first area ever mapped geologically by William Smith (1788). Its location is shown in SW England together with a cross-section.

overlain by lithified sedimentary rock. His classic observation was at Siccar Point, Berwickshire, a couple of day's horse ride south of Edinburgh (see later, Figure 8.1). He could not avoid the conclusion that an immense but unknown time interval was required for the lithification of the lowest sediment, its emergence above sea level, its tilting, its erosion, its submergence and the deposition and lithification of an overlying series of strata. (For this example, we now know the time interval exceeds 70 million years.)

Pragmatically, coal miners in Somerset, SW England (like miners elsewhere in Europe) had already recognized that within a specific region, certain strata always occur in the same order in a given coalfield (Figure 1.2) and they may be correlated from one area to another (Figure 1.3). Moreover, in Somerset, miners knew that certain fossils characterized certain strata. This ordering was not unique to one mine but occurred throughout an extensive coalfield. Smith first comprehended the combined scientific significance of these facts and their practical value. Although he had no formal education beyond grammar school, Smith was a keen observer and had a natural instinct for the scientific method. His deductions were carefully drawn from observation and tested repeatedly, in different areas, until he was comfortable with his deductions. From his initial geological experiences in the Somerset coal mines, he became a very successful self-taught canal engineer. Canals

were essential to England's industrial growth, especially for the economical transport of the heavy coal required in industrial centers. Smith's canal excavations provided continuous exposure through gently inclined strata, from which he was able to confirm the local stratigraphic order of sedimentary rocks. At the same time, he established the world's first collection of chronologically ordered fossils. (In the poverty of later life, Smith was forced to sell the collection for a small sum and then only with the support of influential friends who pressured for its purchase and preservation. The British Museum of Natural History now houses this collection.)

Smith established the essential principles of stratigraphic order:

1. Superposition: younger strata are initially deposited on older ones.
2. Faunal succession: fossils evolve progressively through the stratigraphic column and many species are characteristic of a certain stratigraphic level. However, the concept of evolution was as yet unknown, until Charles Darwin published his book *On the Origin of Species*, in 1859.

Sadly, during their lifetimes, Smith and Hutton did not receive appropriate acknowledgment for their pioneering contributions. However, by 1830 European intellectual society had become sufficiently open minded, largely through

(a) Principle of stratigraphic correlation

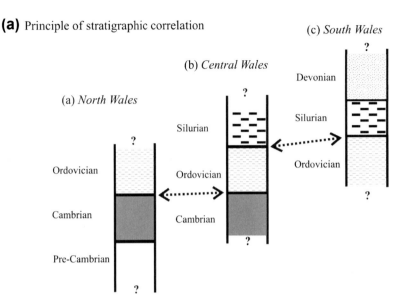

(b) Lithostratigraphic, biostratigraphic and chronostratigraphic correlation shown in cross-section

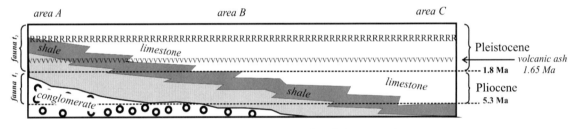

RRRRRRRRRRRRR = geomagnetic polarity reversal
0.78 Ma

FIGURE 1.3   (a) Principle of stratigraphic correlation. Partial sequences in different areas may be overlapped to establish a global stratigraphic column. (b) Lithostratigraphic correlations (=based on rock types) are poor chronological markers and are mostly diachronous as shown here. Chronostratigraphic markers are more valuable, for example, volcanic horizons and geomagnetic reversals that may be assigned absolute ages. Fossils evolve more or less uniformly and also may define chronostratigraphic horizons although they may not be absolutely dated directly.

the effects of the movement known as Scottish Enlightenment, for Charles Lyell to publish *Principles of Geology* in several volumes (1830–1833). This work built on the principles of Smith and Hutton.

UNIFORMITARIANISM (AND ITS LIMITATIONS)

Lyell's books formally established the principles of superposition of strata, of faunal succession and the significance of the unconformity in a manner that was acceptable to the academic community. However, it expanded a further essential concept, the *Principle of Uniformity*, introduced by Hutton that has served geology very well. We may paraphrase uniformitarianism as follows. Any geological process occurring today probably also occurred in the past, therefore, the results of modern geological processes are a key to understanding ancient geological processes. Modern science confirms this as far as erosion, transport,

and deposition of sediment, its lithification to sedimentary rock, and most physical and chemical crustal geological processes including many igneous and metamorphic actions. However, there exist some very important caveats, neglected in most first-year geology courses, which restrict certain processes to certain parts of the geological time scale. Mostly, these concern important differences in processes between Phanerozoic time (the last 600 Ma) and Archean time (anything earlier than 2500 Ma), see Figure 1.4. (The intervening Proterozoic shows a transition in the style of processes.) Three of the major differences that cause exceptions to uniformitarianism are as follows:

1. The Earth's atmosphere was poorly evolved in Archean and early Proterozoic times. Its low oxygen content caused two important differences to more recent time, e.g.,

   a. Iron minerals and iron formations could be deposited as sediment in the reducing atmosphere.

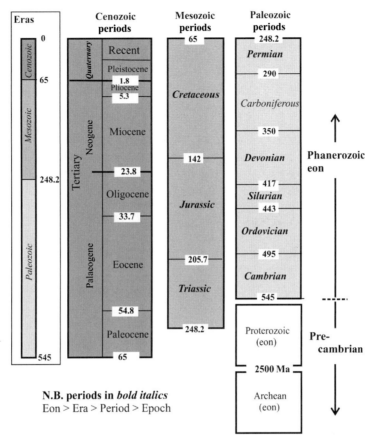

FIGURE 1.4 The relative order of the geological periods was first established in the nineteenth century by stratigraphic correlation of adjacent areas, mainly using fossils. Since approximately 1940, geochronology has provided absolute ages from radioactive isotopes in certain minerals. Absolute ages are defined by and subject to revision by an international commission. As research progresses, new studies bracket the absolute ages of stratigraphic units more precisely and revise boundaries upward to younger ages, in general. For example, when the author was a student, the bases of the Ordovician and Cambrian were pegged, respectively, at 500 and 600 Ma. Over the last few decades, revisions have been a few percent at most.

b. Land vegetation could not survive, thus precipitation runoff was extreme causing faster erosion and coarser sediment.

2. The many complicated processes that selectively separate material from the mantle, to create the crust had less effect in the Archean. Crustal differentiation had produced very little continental crust by the end of the Archean and it was thinner. Thus, high mountains and plate subduction were probably underdeveloped.

3. Ocean crust and ocean dike complexes were absent so that Plate Tectonics, essential to Phanerozoic geology, as we know it did not exist.

4. Probably the most important exception to uniformitarianism concerns the Earth's heat budget. With the exclusion of solar energy driven systems such as the hydrosphere, erosion, and sediment transport, almost all geological processes are driven by geothermal energy. This heat is produced by radioactive decay at depth but, as time passes, the parent isotopes become less abundant. Consequently, in the early Archean, before 3000 Ma, the heat emanating from the Earth was four to eight times higher than now due to the more abundant parent isotopes. This caused the Archean to differ from subsequent history in several ways, including, e.g.,

a. All tectonic process were faster.

b. Volcanism was more abundant and of a different character.

c. Tectonic plates, if they existed would be thinner.

d. Sedimentary formations would be more discontinuous.

e. Certain now rare igneous rocks were more common then (e.g., anorthosites, komatiites).

f. The ways in which rocks deform (fold, fault) was somewhat different; for example, pressure solution was rare and strain rates were higher.

g. Subduction was suppressed so that ocean trenches did not form and high-pressure metamorphism did not occur.

Bearing in mind these caveats, uniformitarianism is still a useful pedagogic tool, and almost 100% correct for surface and near-surface processes of a physical nature. The confidence for this statement lies in the fact that the surface of the Earth had abundant water and was therefore below 100 °C, although ancient geothermal gradients were generally much steeper.

STRATIGRAPHIC CORRELATION

The principle of stratigraphic correlation, deduced by Smith and fortified by the principle of uniformitarianism, relies on the recognition of partial sequences in different areas (Figures 1.3 and 1.5). Characteristic lithologies or fossils permit us to assemble the different partial columns from different regions. In this way, scientists began to establish a universal stratigraphic column, although initially there was no evidence for the actual ages (absolute ages, e.g., 200 Ma) associated with parts of this sequence (Figure 1.3). Of course there are certain caveats; even the novice realizes that over long distances depositional environments may change; whereas sand may be deposited here and now, the sediment may be siltier or muddier there, at the same time. Therefore, this simple lithostratigraphic correlation (correlating simple sedimentary rock types) is somewhat risky, especially over long distances. Special strata such as volcanic ashes or the fine debris produced by meteorite or comet impact are different; they represent a simultaneous sedimentary event over a large area, even over the entire world. For example, ash from Mt St. Helens volcanic eruption spread worldwide in a matter of months ("simultaneous as far as geological time is concerned"). Similarly, a characteristic, worldwide iridium-rich dust layer at the end of the Cretaceous signals a major impact associated with the extinction of the dinosaurs (Figure 1.4).

Strata like this, deposited synchronously over large areas, are termed chronostratigraphic markers. In contrast, lithologies that develop at different times in different areas, for example, due to marine incursion or stream-channel migration produce lithological layers that are not-time markers; they are diachronous (Figure 1.3(b)).

The problems of diachronism or regional inconsistencies in the actual sedimentary characteristic of a formation are overcome with biostratigraphy (Figures 1.3(b) and 1.5(b)). Fossils are characteristic of their time due to the effects of natural selection dictating an evolutionary change. The changes are not gradual in a continuous manner but occur in a staccato fashion, triggered by mutation and environmental stress. In geological terms, the propagation of a new species is usually effectively instantaneous and global. Thus, a certain species will characterize and identify the relative age of the rocks globally. Microfossils that floated around the oceans (e.g., diatoms) satisfy this criterion well; other fossils were usually restricted climatically or by environment so that their global use is somewhat more limited. For further information read the following exciting introduction to the History of Geological Maps and Stratigraphy,

in the form of a popular paperback: S. Winchester (2001), *The Map That Changed the World*. Harper Collins, New York, 329 pp. Fossils, especially microfossils may render an extremely precise specification for ages with individual biozones enduring for less than 5 Ma (Figure 1.5(a)).

More restrictive but equally precise techniques for determining the age of strata include geomagnetic polarity time scale (GPTS) and geochronology of tuffs or lava flows. The GPTS for the limestone strata of Cyprus is shown in a column at the left of diagram, Figure 1.5(b). The geochronological ages are shown in the 60 Ma scale against the Paleogene and Neogene epochs. The contrast between the diachronous lithostratigraphy and the chronostratigraphic markers (GPTS, biostratigraphy, and geochronology) could not be more astonishing; clearly, strata are rarely time markers.

WHAT IS MEANT BY "MAPPING"

In the broadest scientific sense, mapping is the transferring of information from one coordinate system to another. The original information is subject to a consistent mathematical transformation so that it appears in a new format. Usually, the purpose is to show some aspects of the data more clearly since many data sets, especially in the Natural Sciences, have variables that require many dimensions for their precise description. For the geologist, "mapping" is used in a simplified sense; the geologist's map is a copy of nature, which is merely reduced in size and ideally cleared of any spurious information not pertinent to the needs of geology. (I have written "ideally" since some countries simply superimpose geological information on top of existing topographic maps.) The geology map is a plan view, on which boundaries between rock types and geological structures such as faults or joints are shown in their correct location and orientation relative to the geography. The representation is without any distortion, as if seen from a high-altitude satellite photo yet with sufficient resolution to identify the features of interest.

Most geologists map over relatively small areas, of say 10 km × 10 km or less, in order to evaluate some specific geological feature, like the margin of a depositional basin, an ore deposit, a potential dam site or an area with structures that control groundwater or petroleum reserves. For target areas of these sizes, the ratio between the map size and the size of the area is called the representative fraction (RF). Common RF values are 1:10,000 (1 cm on the map = 100 m on the ground) and 1:50,000 (1 cm on the map = 500 m on the ground). For imperial unit maps, there are some quaint RFs, for example, 1:63,360 maps have a scale where 1 inch on the map = 1 mile on the ground. (There are 63,360 inches in every mile.)

This leads to some nontrivial philosophical issues. Suppose the geologist traces out a contact between two rock

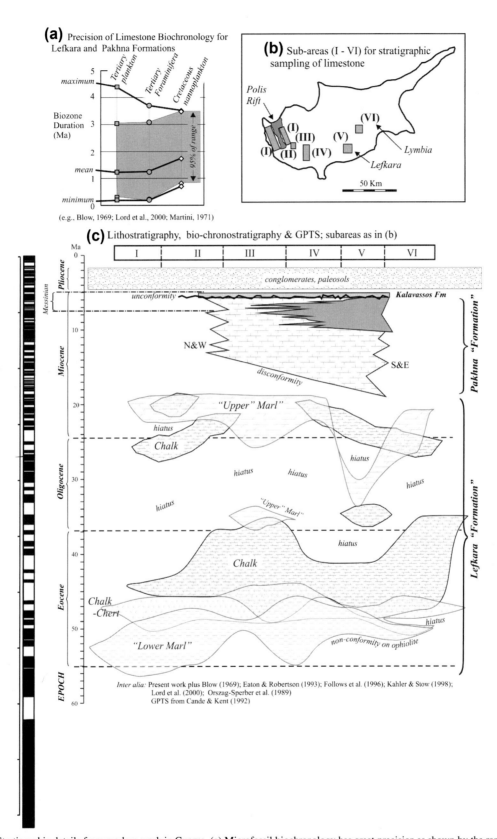

FIGURE 1.5   Stratigraphic details from modern work in Cyprus. (a) Microfossil biochronology has great precision as shown by the range (maximum–minimum) about the mean age. (b) Subareas for sampling used in (c). (c) On the left, the column shows the global geomagnetic polarity time scale (GPTS) (black=normal polarity and white=reversed polarity). Timelines are strictly horizontal, thus in the main diagram, the lithological units are mostly very diachronous. Blank areas indicate nondeposition or submarine erosion.

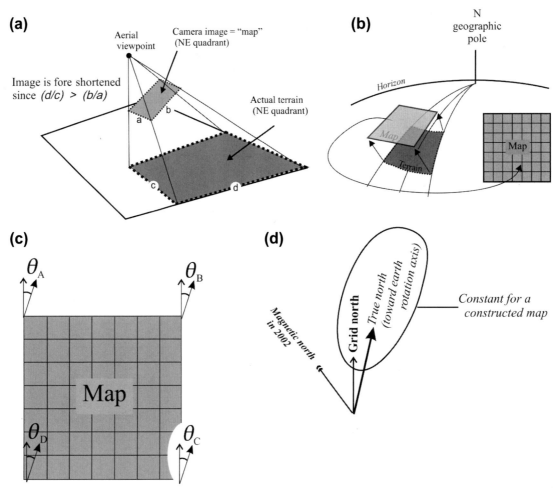

FIGURE 1.6 (a and b) Projection of terrain onto map inevitably causes distortion. (c) The map grid shows true north but magnetic north differs by amounts θ_A, etc., differently at different corners of the map. (d) Magnetic north, to which compasses point, is unstable. For example, for British topographic sheet 09 (an area of 40 × 40 km), magnetic north was 3°38′W of grid north at the center of the map in 2002, changing by 12′E every year.

units in the field, using a 1:10,000 topographic base map (1 cm = 100 m). He carefully records his orientation using a global positioning system unit or simply by visual reference to buildings and fences if they are present on the base map. With care and a 2H pencil, this contact will appear as a line as much as 0.5 mm thick on his base map. The width of his line represents 5 m in the field. Clearly, the average GPS unit is better than adequate to this task. On 1:50,000 maps, the mapped line is now as much as 50 m wide! Key outcrops and specific boundaries cannot be located to better precision than this simply because the resolution of the drawn map will not permit it. Attempting to find the specific locations from published maps may be quite difficult of the vegetation or terrain is inhospitable, since the lines drawn on the map may represent a ground uncertainty of at least ±50 m.

The previous two paragraphs have conveniently considered map scales for which one may assume that the map is a perfect plan view of the ground; there are no problems due to parallax from the viewer or due to the curvature of the Earth. The first problem arises where the geologist uses detailed low-altitude aerial photographs as base maps. Their advantage is that their realism usually permits one to determine a location on the photograph very easily. Whereas the location may be precisely located on the aerial photograph, its location on the Earth is quite another matter. The aerial photograph is a variably distorted ground image (Figure 1.6(a)) and its confusion of topographic variation with geographic position. Very specialized techniques are required to transfer ("map" sensu stricto) information located on the aerial photograph back onto a valid topographic base map. For this reason, as well as the precise location-finding ability afforded by the GPS, low-altitude aerial photographs are now rarely used for geological mapping.

However, high-altitude satellite photographs that are imaged looking perpendicularly at the Earth's surface provide nearly perfect topographic base images. Commonly, these may be found on the internet for the area of your interest but unfortunately images of sufficient resolution to be useful for field mapping are rarely available to civilians.

For regional mapping, the curvature of the Earth becomes a problem. As we learned in high school geography, there can be no universally satisfactory plane map of the Earth's surface. In some way, the surface of the globe or some part of it must be projected onto the plane of the map. Different parts of the map must therefore have different scales. For example, 1 cm may represent 50 km in one part of the map but it may represent 53 km elsewhere. This scale distortion depends also on orientation; the scale is different in different directions. Of course, geological maps of large areas are compiled by assembling adjacent maps of smaller areas so that there is no projection error introduced in their combination. However, to produce a single sheet of plane paper illustrating the geology of many countries or a continent, the original geological maps must be distorted (remapped) into the new coordinate system that cannot avoid scale distortion (Figure 1.6(b)). Most maps on such large scales use some sort of conical projection; the surface of the globe is projected onto the surface of a cone that cuts the Earth in an elliptical path within the area of the map. Any geological or geophysical work that requires quantification must be done with great care since similar distances in nature are represented by different distances in different parts of the map.

At the scales managed by the field distortions due to map projection are detectable although not normally of any practical consequence. For example, the largest area topographic maps used for field geology are usually at a scale of 1:50,000. Maps covering a larger area (say, 1:100,000) on the same sized map sheet do not have sufficient topographic detail to permit us to record geological information and find their location. To facilitate their use, topographic maps of such scales are normally distorted so that they have a square grid. On the 1:50,000 map represented in Figure 1.6(c), 1 cm will always represent 1 km, in every location and in every direction on the map. This is the most practical presentation for geological mapping, we may record distances and simply scale them to represent them faithfully on the map. For the example cited (Figure 1.6(d)), this map covers an area of 40 km × 40 km. Of course, the useful rectangular grid comes at some expense. Wrapping an image of the curved Earth surface onto this square plane involved some distortion. This is most obviously seen by locating the direction to true or geographical north (the location of the Earth's rotation axis). The direction to true north differs smoothly across the map and its extreme values are shown at the four corners (A–D, Figure 1.6(d)). For a map of this scale, the angular distortion is undetectable to the geologist; using a compass, he or she will record directions (azimuths or trends) with a precision no better than ±2°.

This introduces a more vexing question of how the geologist records directions! True north is tied to the map grid but magnetic north is determined by the compass. The magnetic pole does not coincide with the geographic (rotation axis) pole. The magnetic north pole precesses

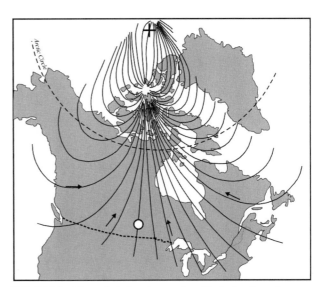

FIGURE 1.7   Magnetic declination mentioned in the last figure is not too extreme for northern Britain. However, for parts of North America, compass needles (heavy arrows) may point far away from geographic north (bold cross at top of map). In the area shown in the white circle, declination is approximately zero and compasses do not need to be adjusted. Elsewhere, the compass's declination screw must be adjusted so that the compass needle points in a meaningful direction, for example, toward grid north or geographic north.

anticlockwise about the rotation axis at very high latitudes; currently, the north magnetic pole is in the Canadian Arctic (Figure 1.7). This Figure shows how extremely compass declination varies across Canada. In less than 1000 years, it will have drifted westward around the top of the Earth to return approximately to its present location. Consequently, to relate directions measured with the magnetic compass to the map, one must adjust the back plate of the compass to correct for the local declination anomaly that is given on the map. Since this changes with time, be careful to use recent maps or make an adjustment for the year if the map and its declination value are very old. If one works with the same compass in areas of different magnetic inclination (especially high latitudes or move to southern/northern hemisphere), it is important to adjust the position of the counterweight on the compass needle.

INITIAL TERMINOLOGY FOR LITHOLOGY

There is much less memory work in this course than in others, emphasis is placed on understanding through practical work. However, read this terminology a few times initially, you will understand it as the course progresses.

Lithology refers to any "rock type", including igneous, sedimentary, and metamorphic rocks. For example, the lithologies in the City of Thunder Bay, Ontario are on deformed Proterozoic (<1800 Ma) shale and diabase, underlain by complexly deformed and metamorphosed Archean (>2700) schist and plutonic rocks. The age of rock is quite unrelated to its complexity and degree of alteration. The Proterozoic shale (~1800 Ma) and fresh diabase (1010 Ma) near Thunder Bay are less altered and fresher in appearance than most Cenozoic (≤65 Ma) shale and diabase in the Mediterranean region. In Cyprus, most Cenozoic diabase has almost none of its original mineralogy preserved; olivine, pyroxene, and bytownite are replaced by chlorite, epidote, calcite, and albite. In contrast, Proterozoic diabase of Thunder Bay completely preserves its original igneous mineralogy.

Lithification refers to complex physical, chemical, or biological processes whereby unconsolidated material (e.g., sand, silt, and mud) becomes converted to solid rock (e.g., sandstone, siltstone, and mudstone); lithos is Greek for rock. Petrification is an alternative Greek word for Lithification but tends to be reserved for the solidification of organic remains by complex chemical processes; for example, fossil wood is often referred to as petrified wood but lithified mud produces a mudstone.

SEDIMENTARY ROCKS

Sediment is deposited on the surface of the solid Earth, on land, or underwater. There are three main kinds of sediment, which form sedimentary rock after lithification.

1. Clastic sediment is most common: rock fragments settle through water, air, or ice. Clast comes from a Greek root for "broken". Sandstone is a typical clastic rock; the fragments of rock may be variably shaped depending on the distance and mode of transport. The clasts settle (sediment out) when the velocity of the transporting medium (most commonly water but also wind or ice) drops below some minimum value. For water-lain clastic rocks, grain size is related to the velocity of the sedimenting water and mineral composition and grain shape are related to the distance of the transport of the clasts.
2. Chemical sediments are deposited by chemical or biochemical reactions, usually underwater; they include iron formations, salts, phosphates, and carbonates. In early Earth history, the low oxygen atmosphere encouraged iron formations; they are common in Archean and Proterozoic history. Desert basins and evaporating lakes are common sites for the cyclical or one-time deposition of salts and phosphates. Carbonates (calcite, dolomite, or ankerite) minerals form the rock limestone and are readily deposited beneath oceans where water temperatures and depths conspire to supersaturate the water with that compound, the carbonate then precipitates.
3. Organic sediments are formed mostly by the accumulation of the hard parts of dead organisms that become fossilized. Most organic sediments are carbonates, formed, e.g., from shells but some are silica based where the remains of deep ocean radiolarian rain onto the seafloor. In the deep marine environment, where there are few

other sources of sediment and sediment may accumulate at rates of a few millimeters per 1000 years, there may be relatively significant components of sediment from the organic remains of phosphate or magnetite (teeth, scales, and bacterial products).

Clastic Sedimentary Rocks

Sandstone	[Sand]
Quartz sandstone	[Quartz sand]
Arkose (feldspar-rich sandstone)	[Feldspathic sand]
Greywacke	[Quartz–feldspar–clay sand]
Flagstone	[Micaceous sand]
Siltstone	[Silt]
Mudstone	[Clay minerals or clay-sized mineral grains]
Limestone	[Calcareous ooze, shell fragments]
Conglomerate	[Pebbles, cobbles, gravel] (pebbles eroded from elsewhere and redeposited at some distance)
Breccia	[Regolith] (angular fragments, eroded, and transported short distances, e.g., talus)

Volcaniclastic Sedimentary Rocks

Volcanic "ash"	[Mineral fragments ejected through air, lain in air or water]
Tuff	Lithified ash
Welded tuff	[Ash fragments "fiamme" melt together]
Volcanic bombs	Larger fragments, streamlined, ejected through air
Ash flow, certain agglomerates	Volcanic ejecta that flow in water or mud from volcanic slopes "epiclastic"
Lapillar tuff	[Lapilli, pellets formed by volcanic dist accreting to raindrops]

There are a few common conventions for the ornaments of certain rock types on maps, however, they are not binding, as you will learn. There are too many rock types for us to reserve a distinctive color, shading, or ornament for each one. Thus, a certain type of shading or stippling may represent tuff on one map but it may represent sandstone on another. Do not confuse terms for sediment with terms for sedimentary rock; in the above lists, sediment terms are enclosed in square brackets.

We do not have space or time to discuss these issues further but it is worth realizing that clastic sedimentary rocks and sediment are described, in the first instance, by their prominent grain size. This is based on grains sizes of unconsolidated sediment that passes through standardized sieves which are custom designed for this purpose. The important and fascinating issues of variation in grain size and mineral content may be cautiously disregarded for the purposes of this book but one cannot proceed in geological studies without a sound footing in mineralogy and petrology.

Classification of Clastic Sediment by Particle Size (Wentworth Scale)

d in millimeters (maximum grain size that passes appropriate sieve)	Characteristic maximum particle size	Description
>2048	Very large	Boulders
>1024	Large	
>512	Medium	
>256	Small	
>128	Large	Cobbles
>64	Small	
>32	Very coarse	Pebbles
>16	coarse	
>8	Medium	
>4	Fine	
>2	Very fine	
>1	Very coarse	Sand
>0.5 (500 μm)	Coarse	
>250 μm	Medium	
>125 μm	Fine	
>62 μm	Very fine	
>31 μm	Very coarse	Silt
>16 μm	Coarse	Clays and muds have similar grain sizes that may comprise microscopic flat-shaped (mica-type) minerals that adhere to one another electrostatically and cannot be separated by sieving
>8 μm	Medium	
>4 μm	Fine	
>2 μm	Very fine	

SUPERFICIAL DEPOSITS (SEDIMENTS AS OPPOSED TO SEDIMENTARY ROCK)

Some sedimentary materials are mostly found in unconsolidated form. For example, glacial sediments, the most common of which are collectively termed till are mostly found as loose material that is mostly young, i.e., Quaternary and due to the recent continental glaciations. Ancient, lithified tills are known from several parts of the stratigraphic column, even from Pre-Cambrian (Proterozoic continental glaciations), but they are exceedingly rare.

Some Glacial Sediments and Features

Till	Poorly sorted material deposited beneath or at terminus of glacier. (Tillite = lithified version.)
Moraine (boulder clay)	Material of widely varying grain size, i.e., poorly sorted, deposited under and around glaciers or by ice rating in periglacial lakes. The range of grain sizes from boulders to clay without internal stratification or order is diagnostic of this glacial deposit.
Loess	Fine windblown silt, eroded from glacial ice.
Varves	Laminated sediment, sometimes graded, sometimes with annual depositional cycles.

Some Special Glacial Landform Feature Recorded on Maps

Drumlin	A landform rather than a lithology, these small hillocks comprise boulder–clay streamlined by ice flow into a tear-drop shape.
Glacial striations	Shown on some glacial geology maps, these are scratches made by glacier-borne rocks, their trend indicates the azimuth of glacier flow. (In the field, it may also be possible to determine the sense or direction of ice flow from the shape of the scratch impression.
Ground moraine	Deposited beneath glaciers, widespread beneath continental glaciers, very poorly drained area with many closed depressions.
End or terminal moraine	Deposited as a ridge along the edge of a stationary glacier.
Lateral moraine	Valley-side debris accumulated along the sides of a mountain valley glacier.
Medial moraine	Formed by the confluence of two lateral moraines as tributary mountain glaciers merge.
Esker	Subglacial river deposit, usually coarse angular pebbles/boulders now recognized as sinuous ridge.

Some Nonglacial Sediments

Colluvium	Loose fine material accumulating at the surface of the Earth, thicker at the foot of slopes. Colluvium is material in the "mass wasting" transport state between terrestrial erosion on topographic slopes or high points and the point at which material enters a stream channel. Every clastic sediment or sedimentary rock passed through the state of being colluvium at one point. It is usually in this stage that most minerals break down ("weather"), possibly to produce completely new minerals that are more stable at the surface of the Earth.
Regolith	Angular fragments of rocky material and of grit or sand, accumulating at the surface, especially in areas of low precipitation, i.e., hot or cold deserts.
Talus	As above, accumulating near the foot of cliffs and mountain slopes.
Alluvium	River (fluvial) deposits.
Lacustrine	Lake deposits.
Soil	A life-supporting mixture of decaying vegetable matter (humus) and colluvium. Its thickness depends on the balance between the rate of disintegration of the underlying bedrock and the rate of erosion of the soil. As a broad generalization, soils are thinner at high altitudes and high latitudes. Tropical soils may be ~100 m thick, temperate climate soils may be 1 m thick. Soil is a dynamic state, an ephemeral accumulation that must be constantly replenished. In temperate climates, most continental rocks may produce a viable agricultural soil in a few hundred years. In most of the Canadian Shield, there is no soil because under those climatic conditions sufficient time has not elapsed since the retreat of the continental glacier to establish a true soil profile. Under the present climatic conditions in the Canadian Shield up to 5000 years may be required too establish a soil profile, dependent on the nature of the local bedrock.

The term "soil" is overused and misused because the layman appears to define it as anything in which plants grow. In fact, plants also grow in any watered natural media that have sufficient chemical elements (~24 elements are required for broad-based agricultural purposes), and from which these elements may be liberated by solution. The latter property is essentially dictated by the fineness of grain size. Nonsoils that permit agriculture include alluvium (every river valley), loess (postglacial windblown material in much of China, parts of North America and northern Asia), and volcanic ash (large parts of SE Asia).

Volcanic ash   Fine solid fragments (mostly <0.5 mm), usually a mixture of crystal fragments and shards of mineral glass, ejected from volcanoes. The fragments are fine enough to settle slowly through the air, somewhat like snow flakes or small hail stones. The ash may be fine enough to circle the world on high-altitude winds but most falls at distances from the volcano controlled by wind direction and strength, and by particle size. Migration patterns of prehistoric agriculturalists tracked successive ash eruptions in SW USA due to the increased fertility following ash falls. "Ash" is another unfortunate misnomer; the material is not the product of combustion but when fresh it does usually posses a gray–black color.

MAP REPRESENTATION OF SUPERFICIAL DEPOSITS

Some governments publish maps that show these superficial deposits prominently because they are so important for mundane human activities (sources of road sand and gravel, construction materials). These may be called "Surficial Geology", "Drift Geology", "Superficial Geology", Quaternary Geology", etc. according to the policy of the Geological Survey. On such maps the bedrock, which is of greater interest to most geologists and throughout this book, may only be ornamented where outcrops actually occur or where the superficial deposits are so thin that the bedrock may be readily uncovered by a shovel. Sometimes, the bedrock will be shown everywhere on surficial geology maps but it may be in a paler color than in the locations where it actually crops out.

Whereas most nonlithified materials are Quaternary, we may find very ancient materials that have escaped lithification. For example, there are Cambrian clays near Moscow and Proterozoic regolith (Gruss) in northern Ontario. On the other hand, Quaternary materials may be very well lithified, for example, in limestone terranes in Southern Spain, regolith is cemented to form a rock hard "surface limestone" called calcrete within a few hundred years.

IGNEOUS ROCKS

Igneous rocks (from a Latin root for "fire", cf. "ignition") are formed by the crystallization of magma, at depth in the Earth. Granite and gabbro are typical examples. Igneous rocks lend themselves to a very rigorous classification based on the proportions of their minerals. However, geologists have alternative names for finer grain sizes. For example, medium-grained gabbro is diabase and fine-grained gabbro is basalt. The smaller the grains size, the more rapid cooling and crystallization.

Typical igneous rocks (fine-grained equivalents in parentheses) are as follows:

Granite   (Rhyolite)
Diorite   (Andesite)
Gabbro   (Basalt)

The fine-grained versions, attributed to rapid cooling, and are typical where magma comes close to the surface of the solid Earth. If the magma erupts onto the surface, it is termed lava.

Consequently, lava is also a stratified rock, although it is not sedimentary! Still more subtle complications occur with explosive volcanicity. This produces showers of energetically expelled crystals and rock fragments which fly through the atmosphere; the finest material may be wind-blown to circumnavigate the globe at high altitude, before it settles. Coarser material (so called "bombs") may be ejected no more than a few kilometers. These materials are collectively termed tuff and commonly misleadingly referred to as "volcanic ash". Tuffs are thus volcaniclastic sediment and stratified in most cases. Where tuffs settle through water, they may superficially resemble normal, water-lain clastic sediments in the field and are sometimes referred to as epiclastic.

The actual form of the body of igneous rock depends largely on the level at which it finally crystallizes. The principal shapes of the most common igneous bodies are as follows:

Dike	[Crosscutting sheet, near surface, not necessarily vertical]
Sill	[Sheet concordant with adjacent rock bodies, not necessarily horizontal]
Apophysis	[Discordant offshoot from another body, such as a sill, dike, or pluton]
Volcanic vent or pipe	[Cylindrical, vertical, reaching surface]
Laccolith	[Dome-shaped body, thin in comparison to areal extent]
Pluton	[Originally deep, large in volume, some plutons in the South America Andes are larger than small European countries such as Wales or Belgium. In some cases, the pluton is an original magma chamber. However, the term is also used for large volumes of metamorphic rock that have similar mineral assemblages to igneous rocks like granite and tonalite, discussed below.]
Intrusion	[Umbrella term for all the above]

METAMORPHIC ROCKS

Metamorphic rocks are sedimentary or igneous rocks that have been changed in mineralogy and texture, usually at great depth due to the effects of any combination of heat, pressure, strain, and aggressive fluid interaction. They are not "melted" rock; the original minerals are partly or completely transformed to new minerals by solid-state diffusion, although this may be aided by diffusion through a fluid (not a melt).

Metamorphism:

1. Increases the density of rocks.
2. Reduces the water content of minerals.
3. Reduces the number of minerals.
4. Introduces minerals with more varied elemental composition.

Metamorphic rocks mostly (exception is contact metamorphism) show an alignment of minerals (due to strain) and the ornaments on maps sometimes show this. Most metamorphic rocks are folded. However, folding is not restricted to metamorphic rocks; many sedimentary rocks fold without any metamorphism, simply by rearranging the positions and orientations of mineral grains, aided by fluid action. Map ornaments often indicate metamorphic rocks with folded layers (bent bedding planes).

Typical metamorphic lithologies with their protoliths [original rock] are as follows:

Slate	[Mudstone]
Pelite	[Mudstone]
Quartzite	[Quartz sandstone]
Greywacke	[Arkose, feldspathic sandstone]
Psammite	[Any sandstone]
Schist	[Slate, mudstone]
Marble	[Limestone]
Greenstone	[Basalt]
Amphibolite	[Basalt or basic tuff]
Paragneiss	[Schist, slate, mudstone]
Orthogneiss	[Some igneous rock]
Gneiss	[Coarse grained of unknown protolith]

Occasionally, the prefix "meta" is added to a protolithology name to indicate that it is now in a metamorphic state, e.g., metagreywacke.

Metamorphism is the most difficult aspect of petrology that a student will meet. A few generalizations will help at this stage. Progressive or prograde regional metamorphism produces

1. denser minerals,
2. fewer minerals (at a maximum six due to the application of Gibb's phase rule),
3. less hydrous minerals (H_2O is always driven out),
4. minerals with more elements (greater solid solution),
5. alignments of minerals (textures or fabrics) related to finite strain or at very high temperatures due to syncrystallization stress,
6. groups of minerals (assemblages) associated with ranges of pressure and temperature (metamorphic facies). As examples, subduction zones are characterized by blueschist facies, orogenic metamorphism on continental margins or at collision zones shows greenschist and amphibolite facies metamorphism, and
7. reversal of metamorphism (regression) to "less metamorphic" assemblages is very rare and incomplete; thus metamorphism invariably records the ultimate change.

Plutonic Rocks

There are many examples, particularly in the Canadian Shield and other Archean terranes (>2500 Ma), where it is difficult, even for an experienced geologist, to decide whether a rock is igneous or metamorphic. Invariably, these rocks are coarse grained, if fossils or sedimentary structures were present, they could not have survived the changes; only rarely does a ghost stratigraphy survive to prove a sedimentary origin. Almost always, any convincing primary sedimentary or igneous character is lost. Although they are commonly banded, the layers may commonly be shown to result from tectonic deformation and metamorphic differentiation. The unique characteristic that distinguishes plutonic rock from sedimentary and igneous ones is the presence of deformational-metamorphic textures, especially preferred lattice orientations of crystals. Even if they are observable in the field, without the benefit of microscope work, it may be difficult to distinguish preferred lattice orientation due to metamorphism from that caused by magmatic flow.

The coarse grain size of plutonic rocks does readily permit a mineralogical classification, however, even in the field. Regardless of whether the rocks had a sedimentary protolith (paragneiss) or an igneous protolith (orthogneiss), we may readily apply an "igneous" label to the mineral assemblage. For example, many such rocks in the Canadian Shield have the same minerals as granite, granodiorite, syenite, or tonalite but advanced studies using petrography, structure, and geochemistry can be prove they were originally sedimentary, subsequently modified by severe metamorphism. To distinguish a plutonic rock from a mineralogically similar igneous one, many geologists would substitute granitoid for of "granite". A purely descriptive term would avoid confusion and would not prejudice future discussion. (The history of geology and biology abounds in the precocious adoption of a genetic label.) Preferred, less biased plutonic rock terms include the following, with the original rock [protolith] indicated in brackets. These are reasonable generalizations:

Granitoid	[Any continent-derived sediment or igneous granitic rock]
Charnockite	[Idem]
Eclogite	[A basic igneous rock, commonly basalt or mafic tuff]
Granulite	[Any continent-derived sediment]
Anorthosite	[Feldspar-rich gneiss of arguable genesis]
Migmatite	[Any clastic sediment, subject to extreme metamorphic differentiation; partial melting could be involved in some cases]
Anatectite	[Quite rare, any clastic sediment or granitic rock, convincingly affected by melting]

Relative Ages

Chapter Outline

In this book, our attention is almost exclusively focused on relative ages; is this fault older than that granite intrusion or is this sandstone older than that conglomerate? From the map alone, we never know how much older or how much younger. That would introduce the question of absolute ages, discussed in the next chapter. Relative ages are very important to the geologist, enabling us to place the rocks and the structures in a sequence that permits us to draw two important types of conclusion from a map view alone:

1. The sequence of geological events (geological history), from a map view alone.
2. The three-dimensional relations, i.e., positions and shapes of the rock formations and structures beneath the surface, from a map view alone.

The latter is of great practical importance; we may estimate useable volumes of economic materials, depths to important features, and the underground extent (subcrop) of rock types (lithologies) of interest. Surprisingly, after training and practice, many of these techniques may be applied largely by visual inspection of the geological map. The principles are very elementary and illustrated on the next page, since diagrams best to explain them (Figure 2.1). However, in brief, the principle techniques of relative age determination include the following.

1. Superposition

Younger strata succeed older strata. This is the normal situation during the formation of rocks (deposition of sediment, extrusion of lava, and settling of crystals in a magma chamber).

However, the tectonic deformation of rocks may fold or thrust rocks so that the order of the original layers is inverted. Special sedimentary features, "way up" indicators

(e.g., cross-bedding, ripples, load casts, burrows, and faunal sequence) may indicate the original top of the bed. Thus, one may confirm that the strata are right way up or upside down. We shall not discuss the exception of inverted strata very much and you may ignore it until perhaps the last chapter of the book. In any case, a good geological map should use a special symbol to warn you where strata are inverted.

2. Crosscutting relationships

Logically, a younger feature should be superimposed upon an older one and therefore cut across it. For example, older strata may be cut by a volcanic pipe, an igneous intrusion, an intrusion of salt, or an igneous dike. The crosscutting relationship should be obvious in most views, in cross-sections or in a view of a cliff but also looking down on the map.

The crosscutting relationship is the fundamental characteristic of most depositional structures that are used to show which way beds young (i.e., which is the top and which is the bottom of the bed). For example, cross-bedding, channeling, ripple marks, worm tubes, grazing trails, mud cracks, syneresis cracks, and sand dikes all rely on the crosscutting principle. These structures are observed in the field by the geologist to determine the stratigraphic order while he is making the geological map.

Note that some care is needed when interpreting relative ages involving minor igneous intrusions such as dikes and sills. Sills and laccoliths (sills with a domed swelling) intrude parallel to the beds or layers of preexisting rocks, exploiting the weakness provided by the contacts between the layers. Although they may appear superficially to conform as layers in sequence, they must in fact be younger than the adjacent strata. For example, the horizontally bedded Proterozoic shale around Thunder Bay, Ontario is about

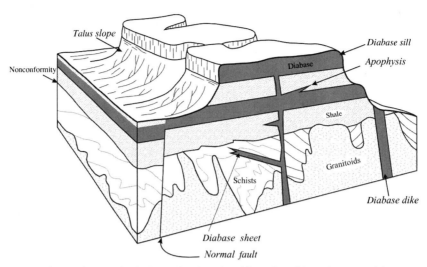

FIGURE 2.1 Block diagram showing relative age criteria. Determine the order of formation of the rock types and features and tabulate it in a list with the oldest at the base.

1900 Ma old. Within this sequence are found thick diabase sills that conform perfectly to bedding but the sills are only 1009 Ma old. Note that a sill is defined as a concordant sheetlike igneous intrusion, whereas a dyke is a discordant (crosscutting) sheetlike intrusion. Most sills dip gently and most dikes dip steeply but that is coincidental and not an essential part of the definition. (There are good mechanical reasons why sills and dikes tend to prefer horizontal and vertical orientations, a subject discussed later.)

One notable exception to the rule of crosscutting relationships occurs with fractures, such as faults and joints, and almost universally with tension joints. A fracture represents a free surface across which it is difficult to transmit stress. Thus after one fracture has formed, it is difficult for a subsequent stress regime to cause a fracture that will cut across the earlier fracture. Consequently, later fractures tend to terminate abruptly against older fractures. This is not universal true, especially for faults with large movements or large extents.

3. Inclusion principle

Under certain circumstances, a rock may include some fragments of older rock. For example, fluvial (river) sand may include pebbles derived from older rock. After lithification, the sand and pebbles together form a conglomerate. The pebbles may be very much older, and of sedimentary, igneous, or metamorphic character, whereas the sandstone matrix portion of the rock constitutes a newly constituted sedimentary rock. The pebbles are inclusions; their age represents the age of their provenance, not the age of the conglomerate. Some inclusions may be almost the same age as the rock that includes them. For example, in some energetic depositional environments, fast stream currents or turbidity currents on continental shelves may rip up contemporaneous

sediment. These fragments may show evidence of their soft unconsolidated character at the time of inclusion, so that is a clue to their young age. However, if the inclusion is lithified (hard rock), it is not possible to determine how much older it is without specialized geochronological work. The degree to which an inclusion has been rounded and shaped during transport is a poor indicator of its greater age.

Terrestrial and submarine landslides commonly pick up other rocks during their movement. Where fragments are abundant these may be described as slump breccias. (Breccia is somewhat like a conglomerate except that the older fragments are obviously not waterworn and they may be angular.)

Igneous rocks commonly contain inclusions. They are termed xenoliths ("foreign rocks"); they may be sedimentary or other pieces of country rock that is picked up at the edge of the magma chamber as the magma intrudes. Volcanic pipes may be choked with xenoliths, debris formed as the volcanic vent forces its way through the crust. Kimberlite pipes may bring diamond-bearing xenoliths from depths >700 km, and many basalt lava flows carry inclusions of the upper mantle to the surface of the Earth. Xenoliths may be large enough to appear on some maps, for example, roof pendants of country rock cover several square kilometers in the Peruvian granite batholiths.

Certain igneous intrusions that contain many fragments of previously lithified rock are termed agglomerates. Not to be confused with conglomerate, agglomerate has an igneous matrix and usually most of the fragments are of igneous rock, which are mostly angular.

4. Faunal succession

Long before Charles Darwin's comprehension of natural selection (1859) guiding the progressive evolution of organisms, humble miners were aware that certain fossils characterized certain sedimentary strata. William Smith recognized that the order in which different fossils appeared in successive strata was the same in different areas. Moreover, whereas the details of the sedimentary rocks changed from one area to another, a certain bed being sandier here, muddier there, the bed would be characterized by the same fossils. Smith grasped the concept that fossils had a chronostratigraphic value that was far more general than lithostratigraphic correlation based solely on the type of sediment. For example, depositional environments may change from one area to another, at the same time. Although the bed deposited at that time may defy lithostratigraphic correlation because its composition changes from sandy at one location to mud elsewhere, its fauna would be the same. Thus, biostratigraphy is a very powerful way of correlating strata of the same age over great differences, in many cases even globally (Figure 2.3).

The change in fauna with time is not gradual but rather staccato in character. Faunal evolution is triggered by mutations, environmental stresses, and mass extinctions that favor some organisms over others. Severe dramatic faunal changes are associated with major climatic or sea level changes. The most dramatic extinction level events are associated with meteor or comet impacts, there are many but the best known is the end-Cretaceous extinction of the dinosaurs. Although the impact site, and evidence there may be difficult to find such impacts cause worldwide dispersion of exotic elements and sometimes a rain of glassy fragments (tektites) at a discrete chronostratigraphic level.

The advent of geochronology has permitted us to estimate the rates of faunal evolution. It is of interest to note that Ammonite fauna evolved sufficiently frequently during the Jurassic and Cretaceous that they provide a resolution of approximately 1.0 Ma. In the parts of the Tertiary, foraminiferal evolution is 10 times more precise; it may permit resolution with a precision of ~100 Ka. The precision of faunal age determinations may be quite good, as indicated for an example of microfossils from Cyprus (Figure 1.5(a)).

An important complication usually found on a large scale and not dealt with in the examples in this book is diachronism. This occurs where a given lithology is deposited at different times in different locations, for example, due to progressive uplift or due to shoreline retreat. The result is that timelines (chronostratigraphy) determined from geochronology of volcanic rocks or magnetochronology determined from field reversals crosscut the lithostratigraphy (rock types). Examples are shown in Figure 1.5 from the limestone sequence of Cyprus, and another well-known example occurs in the North Sea where the 780 Ka magnetic reversal transgresses lithostratigraphy. Magnetic reversals are particularly important since they are globally synchronous phenomena.

5. Chilled contacts in igneous rock

Although this will not be evident at the scale of a geological map, it is an important observation that may determine the sequence of igneous rocks, which, in turn may have been used to construct the legend for a published map. A chilled contact occurs where magma encounters a colder medium. The bulk of the magma cools slowly and has time to form large crystals. However, the margins, in contact with colder rock, cool more rapidly and form finer crystals. Crystallization may even be entirely suppressed and the margins of the magma may form a glass, especially where the magma encounters water or wet sediment. Thus, the relative ages of two igneous rocks may be determined by field observation. As an example, the Proterozoic diabase sills around Thunder Bay took the order of 100 Ka to cool; their temperature was >1300 °C and the adjacent Archean shale was at ~150 °C. The margins of the coarse diabase chilled to form fine basalt in which crystals are not visible to the naked eye. In ophiolites, chilled margins are abundant in dikes and pillow lava.

Chilled margins may also occur for lava flows but, with the exception of submarine lavas, they are usually preserved at the base since weathering and erosion may remove the upper surface. Subterranean intrusions, such as dikes and sills, usually show equally developed chilled margins on all sides since the adjacent rocks are uniformly cool. Exceptions occur in igneous complexes where multiple intrusions occur adjacent to one another in short intervals of time, e.g., ophiolite dike complexes.

Situations which permit relative age determination are illustrated in the three-dimensional block diagram of Figure 2.1; this represents the geology near Thunder Bay. Can you establish the order of rocks and geological events listed from the information in the diagram? Which of the above principles did you use in each case? Include the following rocks, features, and events in the table:

Erosional surface
Diabase dike
Diabase sill
Tilting
Lava
Shale
Metamorphism
Schist
Metaconglomerate
Slate
Folding
Conglomerate

Rock or event youngest at top	Principle or rule used to determine relative age, e.g., superposition and crosscutting relation

In Figure 2.2 a schematic cross-section is drawn which summarizes the principal geological features surrounding near Armstrong, north of Thunder Bay, Ontario.

1. Identify the following features on the map with labels: fault, unconformity sill, dike, sill apophysis, discordant intrusive contact, folded metamorphic rock.
2. Establish a stratigraphic table of events and rocks, with the oldest at the bottom.
3. Geochronological studies show that the age of the diabase is 1010 Ma, the granitoid cooled at 2730 Ma, and the shale was deposited at 1870 Ma. The shale hosted the World's richest silver mines. At what age was the shale most probably mineralized? What is the limiting age for the folding and metamorphism of the schist? A limiting age requires the use of a symbol such as >, <, ≥, or ≤. Where both an upper age limit and a lower age limit for a rock are available, they are said to bracket its possible ages.
4. Which rock type is plutonic and which is igneous?

5. Indicate a sill apophysis on the diagram and a talus slope.
6. What is a limiting age for the fault?

Two entirely hypothetical cross-section puzzles that test your use of relative age determination are shown in Figures 2.3 and 2.4. In each case, you should be able to construct a stratigraphic column of rocks and events, with the oldest event at the base, although some ambiguity is always possible in the order of the sequence of events. Always place the oldest event at the bottom of the table and the youngest at the top; work in pencil since one normally needs to correct and improve the solution as one works through the problem. Remember to include events like tilting, uplift/ erosion in the sequence of events. To the table, add reasonable absolute ages (in millions of years) using Table 1.2.

MORE ADVANCED CONSIDERATIONS IN RELATIVE DATING

Structural and Tectonic Applications

Two more specialized parts of geology use aspects of relative dating that provide critical information that could not otherwise be obtained. The two subjects are structural geology and paleomagnetism. In structural geology, small-scale structures, such as folds, may occur in temporally different episodes associated with different periods of orogeny (mountain building). The episodes of deformation are usually designated D_1, D_2, …, D_n. Originally planar layers may be bent into more-or-less regular waveforms by ductile flow processes, usually accompanied by metamorphism in which new minerals grow in new orientations, controlled by the strain or stress axial orientations. The new minerals usually align to form a fabric such as slaty cleavage or schistosity. In Figure 2.5(a),

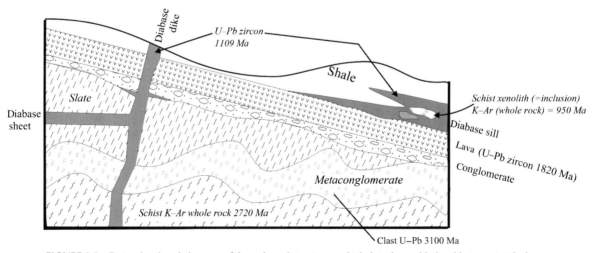

FIGURE 2.2　Determine the relative ages of the rocks and structures and tabulate them with the oldest event at the base.

some first episode folds (F_1) are accompanied by axial–planar S_1 cleavage. In a subsequent deformation event, some or all of the first folds may be refolded to define F_2 folds with an S_2 cleavage. The recognitions of such structures isolate different Earth movements associated with pulses of strain during an orogeny. In one part of SW Scotland, eight different episodes were recognized during the Caledonian orogeny; however, four phases of folding and three phases of cleavage development are ubiquitous through most of the Scottish Highlands. In contrast, in Canadian Achaean terrains, although severely deformed and more strained than most of the Caledonides, only two phases of deformation may be recognized. In outcrop and under the microscope, the relative ages of fabrics are also evident. A first fabric, such as slaty cleavage or schistosity, is pervasive and represents the preferred orientation of micaceous rains or amphiboles throughout the rock. This preferred orientation cannot be readily obliterated by secondary fabrics; therefore, secondary fabrics tend to microfold or crenulated the first fabric. Thus, the appearance of postprimary fabrics is characteristic. At very high metamorphic grade, usually above 550 °C but dependent upon temperature and time available, metamorphic rocks may recrystallize either dynamically or by annealing. The result may be the obliteration of all previous fabrics. Thus, high-grade metamorphic rocks have rather short "memories" of their structural history that poses great problems in their study.

Another common ductile or semiductile minor structure that permits relative age determination is the shear zone (Figure 2.5(b)). This is the ductile equivalent of a fault, although the displacement is usually small; it is commonly identified by a central zone of reduced grain size, dynamic recrystallization, and better developed schistosity. Finally, relative ages of adjacent regional tectonic terranes may be determined by similar techniques. Terranes, characterized by minor structures, metamorphic style, sedimentary facies, faunal age, or geochronology, may be mapped on continental scales (Figure 2.5(c)). The manner in which younger orogenic terranes truncate older ones is reminiscent of the T-junction principle shown by unconformities. Although the orogenic components are already dated absolutely by geochronology in this case, that is not a prerequisite of recognizing the terranes' relative ages from Figure 2.5(c).

A slightly more advanced but still elementary exercise in relative ages appears in Figure 2.6. The best approach to understanding the sequence events in complex cases is to begin with the youngest rock or event, and work backwards in time. This exercise shows several unconformities, revealed by T-junction terminations, discontinuous stratigraphic layers, sedimentation influenced by fault movement, igneous rocks with crosscutting relations, and also showing the inclusion principle of relative dating. More subtle aspects of history may be tackled by discussion with the instructor. For example, how many phases of folding are present? With which phase of folding is the cleavage–schistosity fabric synchronous? Could one infer multiple metamorphic episodes? In which parts of the E–W section which of the following terms would be appropriate for the diabase: dike, sill, concordant sheet, or discordant sheet? To show your understanding of the three-dimensional nature

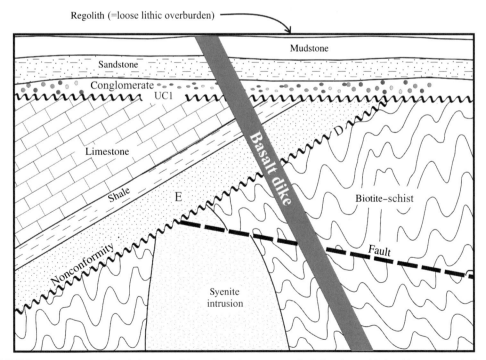

FIGURE 2.3   Determine the relative ages of the rocks and structures and tabulate them with the oldest event at the base.

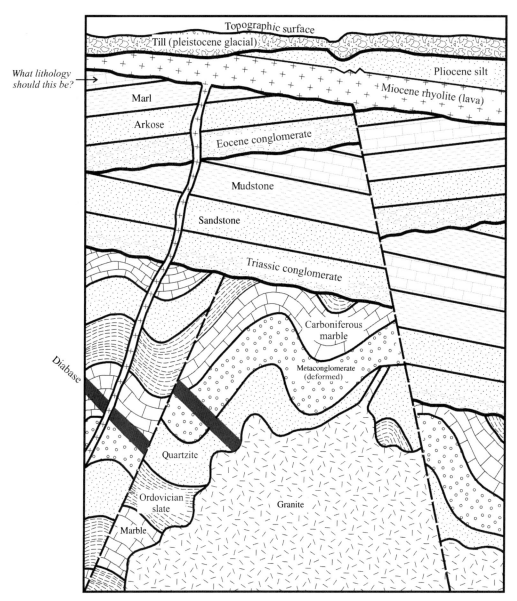

FIGURE 2.4 Construct a table of lithologies (=rock types) and events (tilting, folding, faulting, and intrusion) in ascending order. There may be some ambiguity. In your stratigraphic column, your blocks are symbolic in size and do not represent actual thicknesses or time intervals. Indicate reasonable possible stratigraphic ages (e.g., "Jurassic" with an absolute age in millions of years). Name also two unlithified rock types, two lithifed rock types, two metamorphic rocks, two intrusive igneous rocks and one extrusive (lava) igneous rock type.

of the geology, draw a possible cross-section along the N–S traverse, and then attempt to sketch a possible map at the top of the diagram. Note my use of the word "possible", since there is no unique solution. Discuss this with your instructor.

PALEOMAGNETISM: A SPECIALIZED APPLICATION OF RELATIVE AGES

The goal of most paleomagnetic research is to determine the location of some rocks relative to the ancient North Pole, and relative to some other rocks, from the permanent magnetism of its ferromagnetic minerals. This supposes that the remanent magnetism is ancient and primary, which is not always the case. A certain part of the permanent magnetism is isolated by complicated laboratory experiments, and it is the direction of this magnetic component, the characteristic remanent magnetism (ChRM) that must be associated with the very origin of the rock if paleomagnetism is to be of any use in this exercise. This requires certain specialized aspects of relative dating that present a good mental exercise for us all at this stage, even if paleomagnetism is not in our immediate curriculum.

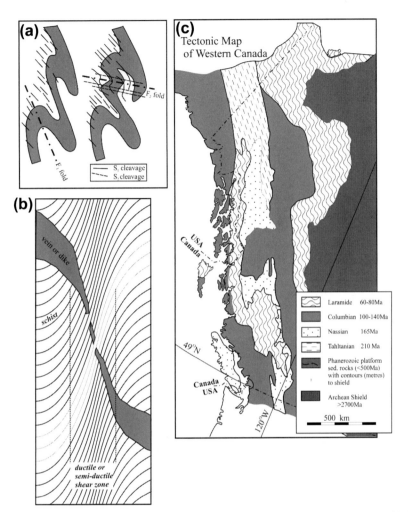

FIGURE 2.5 In metamorphic rocks, structures often reveal relative ages. (a) A first fold is refolded in a second episode of deformation. (b) A shear zone modifies the preexisting schistosity. (c) Tectonic terranes, dated geochronologically, show relative ages where they crosscut one another.

Storage and Masonry Test

We ask ourselves how stable is the ChRM; does it endure for a long enough time for our purposes? If so, it is at least possible that it dates from the formation of the rock. One test is the storage test. One may determine the orientation of the ChRM, then store the specimen either in a fixed orientation in the Earth's magnetic field or inside a magnetic shield. If the ChRM vector is unchanged in direction after several years, it is possible that the magnetism is stable. In a similar theme, if we find that all of the outcrops in an area are magnetized exactly parallel to the present Earth's field, it is unlikely that they carry an ancient magnetism. This would be further corroborated if masonry (stone walls, etc.) constructed of that stone were also magnetized parallel to the Earth's field.

Conglomerate Test

The simplest geological test of stability is the conglomerate test. Suppose that for a lava flow, we determine the ChRM

vector has a trend of 315° (=NW) and a plunge of +20° (positive plunge = angle of down-plunging vector; negative plunge = angle of up-pointing vector). Is this ancient and therefore useful in continental reconstructions?

Higher in the sequence, we find a conglomerate that contains pebbles of lava derived from the lava flow. Let us measure the ChRM direction in the pebbles. If these vectors are scattered, then we may assume that the ChRM was so stable that it survived erosion and sedimentation and endured for millions of years after its erosion. It is reasonable to infer that the ChRM is primary; it does not fail the conglomerate test. Note that there is still no guarantee that the ChRM is primary, so we can never say that the conglomerate test is passed or accepted; only that it is not rejected. This is directly analogous to hypothesis tests in statistics. On the other hand, if the magnetic vectors in the pebbles all pointed to the same direction, we would know that they had remagnetized after the formation of the conglomerate. This result would prove that the ChRM was unstable and unsuitable for paleomagnetic studies.

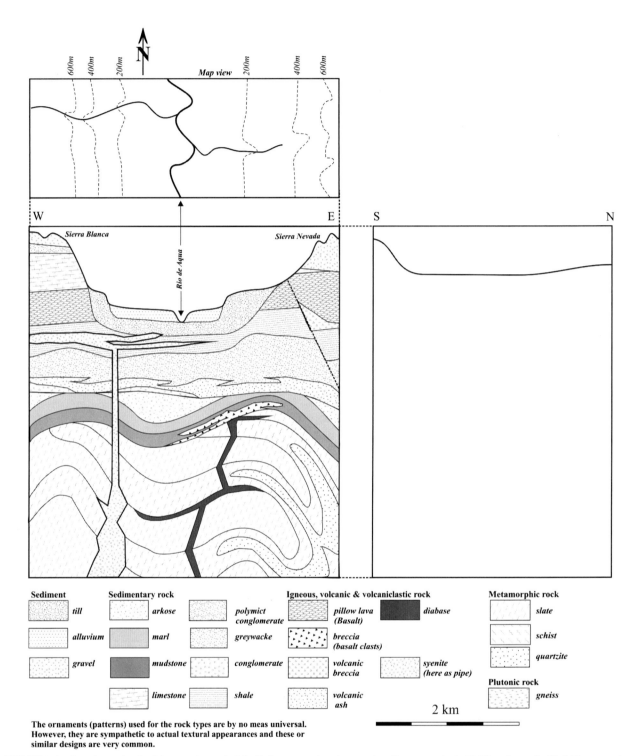

FIGURE 2.6   Complete a sketch of the map view and of the N–S cross-section. Assume that all structures in the E–W section strike N–S. Thus, at a given elevation, each geological formation will trace out an approximate N–S line.

Fold Test ("Tilt Test")

Some folds that have gentle and uniform dips to their flanks essential rotate any feature in the limbs in a rigid-body manner. This means that it is possible to geometrically restore the flanks of the fold, and any features in them to their original attitude. Note that most folds do not satisfy these constraints because they involve shape change (strain). Whereas paleomagnetists informally refer to "unfolding" the limbs of the fold, they are in fact merely untilting the layers and the restoration is only accurate if the fold formed without any strain (shape change) of its flanks; this is only approximately true in very open angular folds.

Consider one of the rare folds suitable for our purpose; if the paleomagnetism is stable in the beds, the ChRM direction will have been tilted precisely in unison as the beds steepened in to their tilted orientation. We may now geometrically "untilt" the beds on each flank until they coplanar and their ChRM vector directions from each flank should coincide. If the untilting procedure does not produce similar ChRM directions in the two flanks then the fold test is failed. In that case, the paleomagnetism is not ancient enough to be associated with the actual formation of the beds.

Baked Contact Test

Consider an intrusion into country rock that has its own stable remanent magnetism. When the intrusion cools, it usually acquires a strong remanent magnetism but its heat will also remagnetize the country rock in the ambient field direction at the time of intrusion. Thus, if we find that our intrusion has the same ChRM direction as the baked contact, we do not reject the hypothesis that the ChRM is stable.

Reversal Test

The magnetic field lines of the present geomagnetic field point toward the magnetic north pole which is currently located in northern Canada. The actual magnetic poles never coincide with the geographic poles (=where Earth rotation axis pierces Earth) but this is only a major problem for the use of a magnetic compass when working at very high latitudes. The present north-seeking behavior of compasses has persisted for ~780,000 years (The Brunhes' Epoch of Normal Polarity). Before that, compasses pointed toward the south part of the Earth. Moreover, the geomagnetic field has alternated between normal polarity (as now) and reversed polarity throughout geological history. With the exception of some quiet intervals, and ancient parts of the geological column with insufficient data, we now know that reversals have occurred approximately every million years or so. Clearly, if all the rocks of the World had very short magnetic memories, we would not know that there had ever been field reversals; all the magnetism in rocks would point northward rather like the average of all compasses. However, in thick sequences of many lava flows, especially in places like Iceland, Northern Ireland, or Hawaii, the stratigraphic column shows groups of flows with alternating magnetic polarity. This verifies that their magnetism is primary. The pattern of reversals forms a magnetostratigraphy, represented by a modified "stratigraphic column" that shows alternating normal polarity (indicated by black) and reversed polarity (white). The fact that reversals exist shows that the magnetism is primary. The duration of these polarity intervals has been documented by palaeontology and geochronology so that their time series patterns not only provide a correlation tool; their calibration with absolute ages permits them to be used to date rocks that are difficult to date absolutely by other methods. For example, in NW Europe, the migration of early man has dated, by comparing the magnetic directions in strata bearing evidence of hominids with the known magnetic polarity time scale.

Magnetostratigraphy is also prominently displayed by igneous dikes intruded at the spreading centers of oceans. Their remanent magnetism is so strong that its magnitude and direction may be measured from aircraft. (Knowledge of the background ocean floor magnetism if one wishes to detect the presence of submarines from the air by magnetometry. It was this military application that first lead to the discovery of ocean floor spreading and thus of plate tectonics.) As the ocean floor spreads by gravitational gliding from the mid-ocean ridge (MOR), batches of nearly vertical dikes intrude parallel to the rift, following the center of the MOR. These dikes magnetize shortly after sea floor hydrothermal metamorphism, which follows igneous cooling; at this time the dikes magnetize in the direction of the contemporary geomagnetic field. Each part of the ocean spreads from the MOR at a rate commonly $0 \geq 5$ cm/year or $0 \geq 500$ m/Ma. Since the Earth's field changes polarity almost every million years, the ocean floor is magnetized with alternating polarity (north-seeking versus south-seeking magnetism) about every kilometer or so, in stripes parallel to the MOR. The stripes are neither as sharp nor as closely spaced as one might hope since the magnetization is not "thermal" following igneous cooling. Instead, the magnetization is due to crystallization and recrystallization following hydrothermal ocean floor metamorphism. This process has fuzzy boundaries and extends much further from the MOR than the "hot" dikes. Most Physical Geology textbooks carry good color maps of the ocean floor magnetic striping; its origin and nature have been confirmed where ocean floor rocks are rarely exposed on land as ophiolites. The ocean floor polarity striping may be regarded as a special kind of stratigraphic column, laid out sideways and symmetrically on both sides of the MOR.

The use of magnetostratigraphy is obviously helpful in volcanic and igneous rocks that cannot contain fossils and which cannot readily be grouped according to a stratigraphic or depositional scheme like sedimentary rocks. At this stage, the reader may be interested to know that some magnetic minerals have the potential to retain magnetic orientations for billions of years and the most abundant remanence-bearing minerals (ferromagnetic minerals like magnetite and hematite) have the ability to retain magnetic field directions from Archean. On the other hand, more refined uses of magnetostratigraphy, use secular variation of the geomagnetic field. This is caused by the rotation of the axis of the magnetic field westwards around the Earth's rotation axis, about once every millennium. The resulting detailed variations in field direction at a given site provide good correlation and dating tools in archeology, hominid stratigraphy, and Quaternary studies.

Figure 2.7 provides a hypothetical cross-section. Tabulate the geological history on the right-hand side of the page, starting with the oldest event or rock type at the bottom.

Figure 2.8 presents a simplified map of the Marathon igneous complex in northern Ontario. Label each of the mapped rock types in chronological order (1 = oldest). Note that three igneous ring complexes are present.

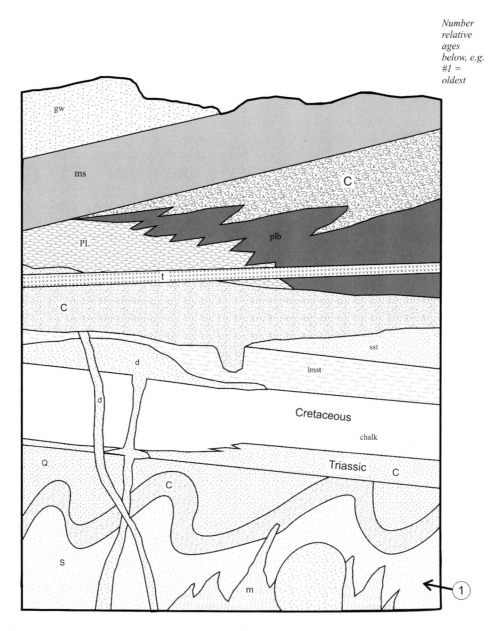

Number relative ages below, e.g. #1 = oldest

gw = greywacke   ms = mudstone   C=conglomerate(s)   plb=pillow lava breccia (hyaloclastite)
PL = pillow lava   t = tuff   sst = sandstone   lmst = limestone
metamorphic: Q = quartzite   S = schist   m = migmatite

FIGURE 2.7   Complete a table of events and rock types on the right-hand margin of the page.

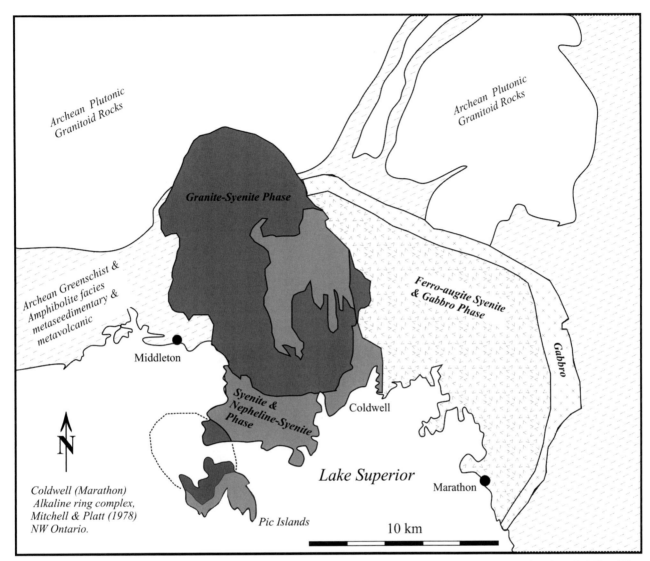

Archean Plutonic
Granitoid Rocks

Archean Plutonic
Granitoid Rocks

Granite-Syenite Phase

Archean Greenschist &
Amphibolite facies
metaseedimentary &
metavolcanic

Ferro-augite Syenite
& Gabbro Phase

Gabbro

Middleton

Syenite &
Nepheline-Syenite
Phase

Coldwell

N

Coldwell (Marathon)
Alkaline ring complex,
Mitchell & Platt (1978)
NW Ontario.

Lake Superior

Marathon

Pic Islands

10 km

FIGURE 2.8 This map illustrates a series of igneous intrusions into plutonic and metamorphic rocks in northwestern Ontario (after Mitchell and Platt, 1978). Place a number (1 = oldest) on each rock type according to its relative age.

Absolute Ages

Chapter Outline

This book requires some appreciation of absolute ages but this chapter is largely parenthetical to the main scope of relative ages as used in map interpretation. However, when working on the map exercises, it is useful to include absolute ages in the description of the geological history, wherever this is possible.

Early natural historians, working in the eighteenth and nineteenth centuries, inferred that large time intervals were required for the formation of rocks, their uplift, erosion, and redeposition as new sediment. Early arguments to quantify the magnitude of geological ages from physical characteristics produced age estimates that we now know to be very low. However, even inaccurate early estimates based on the cooling of the Earth or the salinity of the oceans estimated ages in many millions of years. Earth scientists mostly use informal time units of thousands of years (Ka), millions of years (Ma = 10^6 years) or billions of years (Ga = 1000 Ma = 10^9 years).

UNITS OF MEASURE

For convenience, geologists use "Ma" (million years) or "Ka" (thousand years) as their most common time unit but we must use the second (s) as out time unit in any fundamental calculations. The second (s), kilogram (kg), and meter (m) are the fundamental units of time, mass, and length (SI or Système Internationale units) from which all other magnitude units are derived (e.g., for pressure, stress, viscosity, velocity, and acceleration). It is useful to note that

$$\text{One year} = 3.156 \times 10^7 \text{ s,}$$

$$1 \text{ Ma} \simeq 3 \times 10^{13} \text{ s.}$$

For more advanced work, it may be convenient to use Bayly's (1992) nonfundamental but rather convenient geological time unit, 1 gtu = 10^{14} s = 3.3 Ma. Upon first sight, the gtu is a strange multiple of the fundamental time unit (second). However, it greatly simplifies calculations in structural geology and tectonics since many rocks deform by a very small fraction per second, ranging from 10^{-5} (=0.001%) per second to ~10^{-14} per second. Moreover, 1 gtu ≈ 3.3 Ma which is convenient for managing the huge ages and time intervals required in stratigraphy and geology. Since masses and volumes are so enormous in geological considerations, some other nonfundamental units have been widely adopted to simplify writing. For example, the pressure due to 3 km of overlying rock is approximately 100 megapascals or 10^8 Pa; this is the number and unit that must be used in calculations, along with the other fundamental units based on the meter–kilogram–second system. However, for convenience, the pressure beneath 3 km of rock is almost invariably translated to approximately 1 kbar (1 bar = atmospheric pressure).

EARLY ESTIMATES OF ABSOLUTE GEOLOGICAL AGES

One early analysis used oceanic salinity to estimate the age of the oceans, clearly a very early feature of the Earth. Salt concentrations in "freshwater" streams supply the salts by erosion to the oceans. If there are no other considerations, from the rate of supply of salt and the volume of salt in the oceans, we may simply calculate the age of the oceans. This calculation suggests an age of several hundred million years for the oceans. We now know this is far too young (the oldest marine sedimentary rocks are several billion years old), so what is wrong with this calculation? The salt concentration in the ocean depends on temperature and overall chemistry; excess salt is removed by precipitation. Thus, the "age" calculated in this basis is not the age of the ocean but the mean residence time for salt in solution. That is valuable knowledge in itself but it provides a very low minimum age estimate for the Earth at less than 100 Ma!

Another physical estimate of a geological age concerns the temperature of the Earth. Evidence from meteorites, the density distribution in the Earth and the kinds of geological processes that were more abundant in the older rocks all suggested that the early Earth was very hot. Accepting some primordial molten-gaseous form for the Earth and knowing the temperatures of such siliceous materials permits the cooling age to be determined. From the initial estimated temperature, using the fundamental law of cooling by radiation, a molten silicate Earth surface could cool enough to support an aqueous surface in a few hundred million years. Again, we know this estimate to be far too low. While the information still has value, it assumes that all heat was primordial. The discovery of radioactivity revealed a heat source that could have kept the Earth hot, long after the loss of its primordial heat. Radioactive heat production is still sufficient to drive the motion of the lithospheric plates at speeds of several centimeters per year. Our best modern radiometric determinations now suggest that the Earth is at least 4.54 Ga old.

RADIOACTIVE DECAY AND GEOCHRONOLOGY

For just over 100 years, we have known that certain elements are unstable and decay spontaneously. The rate of decay for a given radioactive element is constant, it cannot be controlled and it is independent of the physical and chemical environment since it is a nuclear process. Radioactive decay produces one or more prominent new daughter elements some of which may also be radioactive. Their mass very nearly equals that of the parent but a small mass of neutrons, electrons, and copious energy is released in the form of X-rays. The associated heat is highly significant in geology, being responsible for the current high temperature of the Earth, as we shall see later. Very soon after the discovery of radioactivity, it was realized that its constant rate of decay and its insulation from other physical and chemical processes made it a good potential clock for measuring the passage of long time intervals.

Several textbooks are available on methods of absolute age determination but most are very technical and some are restricted to one or just a few methods using particular radioactive elements. A useful and readable alternative general resource for both geological and archaeological education is found in:

M. J. Aitken, 1994. *Science-Based Dating in Archaeology*. Longman, Harlow Essex, 274 pp.

Absolute age refers to a quantified age such as 230 Ma or 150 Ka. This can only be determined by complicated and expensive laboratory experiments that measure the age recorded by radioactive decay. The cost of each age determination varies from several hundred to several thousand dollars, depending on the method. An example of a geochronological age determination tool is given by the decay of potassium-40 (^{40}K) to argon-40 (^{40}A). Being a nuclear process, the rate is unaffected by environmental variables such as geological temperatures, pressures, and chemical reactions. Radioactive decay is a probabilistic process so that for the large number of parent atoms remaining in any specimen, the percentage of that decay in 1 s is constant. For example, after infinite time all ^{40}K atoms in a mineral could be replaced by Argon. Minerals have fixed proportions of elements, controlled by their crystallographic structure. For example, there is one atom of K in a newly created orthoclase feldspar [$KAlSi_3O_8$] unit cell. Given the several million (N_0) unit cells in a real mineral, after some time (t, s), ^{40}K will be partially replaced so that only N of the original N_0 atoms remain. The ratio N/N_0 is determined in a mass spectrometer that determines the ratio on the basis of their slightly different masses, using specially prepared samples of the mineral. Since we know the rate of decay (λ), we calculate the duration of the decay process (t), i.e., the age of the mineral, from the following formulas that describes the exponential radioactive decay of the parent element:

$$N = N_0 \, e^{-\lambda t}.$$

Thus,

$$t = -\frac{\ln N}{\lambda \ln N_0}.$$

A useful concept is the half-life; the time elapsed after which half of the existing (not necessarily original) mass remains ($N/N_0 = \frac{1}{2}$).

It is $t_{\frac{1}{2}} = \dfrac{\ln (2/1)}{\lambda} = \dfrac{0.693}{\lambda}$. Figure 3.1 illustrates the decay of the parent isotope (from N_0 units) in terms of half-lives; it can be seen that after five half-lives, a diminished but still manageable quantity remains to be measured.

For ^{40}K, $t_{0.5} = 1248$ Ma, which occur in common potassium-bearing minerals such as alkali feldspars and micas that are useful for dating ancient igneous, metamorphic, and plutonic rocks. After 1248 Ma, half of the parent remains, after 2×1248 Ma, one-quarter remains, and so on. [Note that this is an incremental effect, like compound interest on your bank account, but referring to a decrease rather than an increase.] Clearly, the half-life means that measurable proportions of parent and daughter isotopes remain over

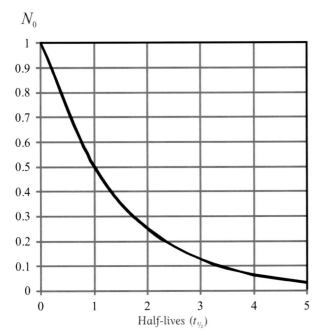

FIGURE 3.1 Exponential decay of a radioactive isotope, measured in half-lives.

several half-lives so that ages of several billion years may be determined by this decay series. But what controls the dating of young minerals and events? Let us consider how long it takes for the existing mass of parent to be reduced by 1%, i.e., 99% or 0.99 of N_0 would remain. The fraction 1% seems to be small yet sufficiently large to be measured accurately in the laboratory. The equation becomes

$$t_{0.99} = \frac{\ln (100/99)}{\lambda} = \frac{0.00436}{\lambda}.$$

For ^{40}K, $t_{0.99} \approx 8\,Ma$. Therefore, without going into a detailed consideration, it is not reasonable to use that particular elemental decay series to determine ages less than several tens of millions of years. This also gives us some idea of the precision of the method; early in the decay sequence, a decay of 1% requires about 8 Ma so that precision is rather poor (±10 Ma). This cautious view is rather pessimistic since the most elaborate geochronological methods use the Concordia technique in which two concurrent ($^{235}U-^{207}Pb$ and $^{238}U-^{208}Pb$) radiogenic series are simultaneously measured. It yields precisions of much better than 1% even for very ancient rocks (>3000 Ma) but its discussion is beyond the needs of this book. Due to the long half-lives of the elements concerned, it is most successful when applied to very ancient rocks.(Table3.1).

Some important but simple considerations are necessary to interpret a radiometric age

1. t is the age since the closure of the mineral lattice to loss, by diffusion, of the daughter element. For example, this might give a meaningful age for an igneous rock

in which the mineral had crystallized. The age would then be the time from which the mineral cooled sufficiently that the lattice could retain the daughter element, which is usually more mobile than its parent. Of course, different minerals become closed systems at different temperatures, so they give different ages even for the same rock. This is particularly noticeable in metamorphic and plutonic rocks that commonly take >10 million years to cool from the temperature at which the new mineral was created. Some radioactive decay systems are more robust than others. However, where the daughter product is mobile as in the K–A example above, loss of the daughter Argon atoms due to spurious geological events would produce false ages that were too young. To avoid minimum ages due to daughter loss, decay series involving heavy atoms are preferred, e.g., $^{235}U-^{207}Pb$ and $^{238}U-^{208}Pb$ decay series. Minimum ages are best avoided if the elements concerned are relatively chemically inactive, immobile with regard to diffusion, and favor sites in minerals that are refractory and chemically stable. The actual radioactive decay is of course a nuclear process and therefore independent of chemical reactions.

2. The radiometric age actually refers to the time of closure of the mineral with respect to daughter loss. As noted above, this is normally soon after the formation of the mineral. However, in some cases, radiometric ages are not used to date the creation of the minerals/rock but events that cause the loss of the daughter element. For example, the age at which a rock was heated by an igneous intrusion or by regional metamorphism or meteorite impact may be dated in this way and the age of regional metamorphism may be determined similarly (Figure 3.5(a) and (b)).

3. One must also choose a decay series that has a half-life similar to the age of the rocks to achieve the best precision. When reading scientific papers, be careful that the rate of decay of the element was suitable for the age, it was intended to determine. We have seen that the half-life ($t_{0.5}$), after which 50% of the parent isotope remains and the 99% life ($t_{0.99}$) after which 99% of the element remains are useful guides. A more useful view of decay series is to compare the 95 and 5% lives, respectively, these yielding a range between the youngest and the oldest events or minerals that may be dated reliably with that radioactive decay series.

4. For sedimentary rocks, almost all minerals are derived from other, older sources. Thus, for fundamental reason, almost every age from a sedimentary rock is a minimum age. The age is actually for minerals derived from an older source rock. Rarely, minerals like glauconite that grow during or shortly after sedimentation (diagenetic minerals) may be suitable for determining the age of sedimentation.

TABLE 3.1 Some Radioactive Isotopes Used in Dating Rocks and Geological Processes

Series	$t_{0.5}$ (half-life) 50% of parent remains after	$t_{0.99}$ 99% of parent remains after	$t_{0.95}$ 95% of parent remains after	$t_{0.05}$ 5% of parent remains after
$^{235}U-^{207}Pb$	4470 Ma	64.8	330.8	19,319
$^{238}U-^{208}Pb$	14,000 Ma	203.0	1036.0	60,507
$^{40}K-^{40}Ar$	1248 Ma	18.6	94.7	5532
$^{234}U-^{230}Th$	75 Ka	1.1	5.6	324
^{10}Be	1.6 Ka	0.02	0.12	6.92
^{26}Al	730 a	10.6	54.0	3155
^{36}Cl	300 a	4.3	22.2	1297
^{41}Ca	100 a	1.4	7.4	432
^{210}Pb	22 a	0.3	1.6	95
^{14}C	5730 a	Not really relevant since reservoir concentration of ^{14}C changes constantly. Actual age determinations require calibration by historical dates, dendrochronology, etc.		

Note: different time units: Ma, Ka, and years (a).

5. Due to the problem mentioned in the preceding equation, the age of sedimentary rocks, and the absolute ages defining the divisions of the stratigraphic column may only be easily determined by two methods:

   a. From the ages of volcanic horizons. These have the necessary properties of igneous rocks but they are stratified within the sequence of beds. Therefore, they provide an age for a given horizon. Lavas, volcanic ashes (tuffs), and other volcanic exhalations satisfy this criterion.

   b. By bracketing the age of a stratum with ages of igneous rocks. For example, sandstone may lie on an erosion surface (nonconformity) above granite. The granite must be older (law of superposition) and its radiometric age gives a maximum age for the sandstone. The sandstone (and the granite) may be cut by an igneous porphyry dike. The dike is therefore younger (crosscutting relations) and its age gives a minimum age for the sandstone. The maximum and minimum ages bracket the depositional age of the sandstone.

A consequence of (a) and (b) is that with the accumulation of knowledge, scientists progressively redefine the ages of key stratigraphic horizons (e.g., the base of the Carboniferous period) (Table 3.2). It is logical that geological research gradually may only find younger critical ages so that the base ages defined by the World International Stratigraphic Commission decrease as knowledge accumulates.

In practice, we now determine radiometric ages from many different radiogenic series. This has the advantage that certain elements decay at very slow rates, useful for dating very ancient rocks, whereas other radioactive elements decay more slowly, providing better resolution for younger rocks and events. Moreover, laboratory techniques are more precise if either the daughter or the parent element remains as a reasonable proportion (e.g., 5%) in the rock sampled. Five half-lives $[=(\frac{1}{2})^5]$ reduce the proportion of the parent atom to 1.6%. Clearly, this is already rather a small proportion to measure precisely. The nature of the sample is also critical; the mineral should have remained a closed system, retaining the entire decay product, otherwise the determined age will be too low. Thus, refractory minerals are preferred; those which are insensitive to subsequent thermal and chemical changes which could permit daughter atoms to escape.

Whereas radiometric age determination is technically very difficult and expensive, its results are more easily evaluated and applied than any other absolute age determination method. This is because it provides a monotonic relationship between the measurement (isotopic ratio N/N_0) and the time t. That is to say, there is a unique age for a given experimentally determined value, and those values change progressively with time (Figure 3.4). N/N_0 changes decreases exponentially with time. Moreover, for each decay series, the decay rate is a universal constant, unaffected by any physical, chemical, or geological process since the decay is a nuclear process.

TABLE 3.2 Worldwide Stratigraphic Column and International Commission Period Base Ages as Accepted in 2002 and 1985

Era	Period	≥Ma (2002) base	≥Ma (1985) base
Cenozoic	Neogene and Paleogene	65	65
Mesozoic	Cretaceous	142	140
	Jurassic	206	195
	Triassic	248	230
Paleozoic	Permian	290	280
	Carboniferous	350	345
	Devonian	417	395
	Silurian	443	445
	Ordovician	495	510
	Cambrian	545	570
Pre-Cambrian	Proterozoic and Late Archean	>545	>570

Fission-track dating is also a monotonic technique, made possible because certain radioactive elements, especially ^{238}U produce microscopically visible damage in certain minerals. The abundance of the fission tracks per ^{238}U atom, measured simply by microscopic observation, is proportional to the age in a monotonic fashion. However, the relationship is not a universal one and requires calibration for a given mineral and radioactive element using conventional geochronology. The tracks are obliterated relatively easily by younger reheating events, thus "resetting the clock".

In contrast, some other well-known absolute age determination techniques depend on some nonmonotonic behavior with time. This means that for some experimentally determined value, there is not a unique age but several, or even very many, possibilities. This usually requires that the experimental observation must be compared with some master curve or calibration curve of values to eliminate the spurious ages and determine which age or ages are reasonable.

The master curve is established by some other dating technique based on values that should have a monotonic relationship with time. This service is normally provided by the laboratory that performs the experimental work. Some age determination methods which depend on measured values that oscillate with time are discussed below.

DENDROCHRONOLOGY

Tree growth rings vary in thickness according to climatic variations and, accumulating at the rate of one per year, they provide a chronological record. With a careful database one may match the pattern of ring thicknesses and thus determine the age of wood included in sediment or in archaeological structures. Obviously, the age is that when the tree fell which gives a maximum age for the sediment or building. Individual trees alive today may be as old as 7000a (Bristlecone pines, New Mexico), 3000a (Sequoias, western North America), 1500a (English yew), and 800a (English oak). Unfortunately, some long-lived species are unsuitable for various reasons, including the idiosyncratic behavior of individual organisms to due to the local microclimate. This is most unfortunate for the case of the English yew, which is found in almost every ancient English and Scottish churchyard, with associated historical documentation. Similarly, olive wood, so common in classical historical sites of the Mediterranean, responds inconsistently to for both dendrochronology and radiocarbon dating. Nevertheless, dendrochronology of successful species like European oak and American bristlecone pine provides a continuous chronological record of approximately 7000a, compiled from overlapping data sets. Climatic variations and localization of species prevent a worldwide dendrochronological record.

VARVE CHRONOLOGY

Deposition in lakes near the margins of continental glaciers may produce laminated fine sediment, with annual increments corresponding to melt rate-controlled deposition. Sections of varved clays quite commonly preserve as many as 15,000 varves (varve=Swedish for cycle) in a continuous record, e.g., in Sweden and in Minnesota. Their thickness variation is precisely related to climate, with much less idiosyncrasy than tree rings. It is even possible to recognize the sunspot cycle (every 11years) when increased solar radiation produces more melting and slightly thicker varves. Varve thickness curves thus prove a useful chronological scale.

RADIOCARBON AGE DETERMINATION

Although this is probably the most familiar scientific dating method for the general public, its complications and limitations are grossly underestimated, even by the scientists and archaeologists who use it. Carbon is a building block of all organic molecules and is used by organism regardless of whether it is ^{14}C, ^{13}C, or ^{12}C. Of these three isotopes, ^{14}C is radioactive and decays with a half-life of ≈5730a to ^{14}N. This represents a compounding decrease of about 1% every 83years. For every million, million C-atoms (10^{12}), there is one radioactive ^{14}C atom. When first proposed by Libby (1955), it appeared that scientists optimistically considered, there was a similar monotonic decay for carbon materials of any age. That is, a 1000-year-old piece of wood was necessarily expected to have less radioactive ^{14}C than a piece of wood 700years old.

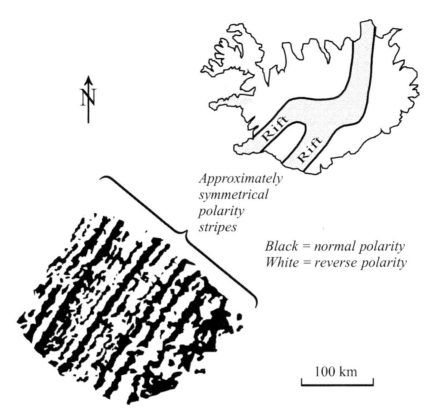

Approximately
symmetrical
polarity
stripes

Black = normal polarity
White = reverse polarity

100 km

FIGURE 3.2 Seafloor magnetic anomalies, SW of Iceland. The black stripes represent seafloor magnetized during normal polarity of the Earth's magnetic field. Note that the stripes are approximately symmetrical on either side of the central Atlantic rift. The striping forms the basis of the global geomagnetic polarity time scale (GPTS).

Unfortunately, the loss of ^{14}C by decay is only approximately balanced by its production by cosmic radiation in the atmosphere due to variations in solar flux and in the strength of the geomagnetic field. Especially over short periods, comparable to the life spans of an organism, the balance is imperfect. Moreover, whereas the carbon isotopes behave identically in biochemical reactions, their different masses cause their concentrations to vary in different species and in the same species at different times. This is primarily a thermokinetic effect; heavier isotopes are less mobile and less readily absorbed in cooler temperatures. Thus, we do not know the fixed starting proportion for ^{14}C. For example, at the same time, ^{14}C will be more abundant in cold deep oceanic waters and less abundant in shallow warmer water. Thus, organisms consuming food from deep water will have skeletons richer in ^{14}C than those from creatures consuming shallow water fish. Contemporary skeletons will thus appear to have very different ages because they drew on reservoirs with different ^{14}C concentrations. Many further complications are discussed, e.g., by M. J. Aitken, 1994, *Science-Based Dating in Archaeology*, Longman, London and New York, 274 pp.

The consequence is that the measured ^{14}C/^{12}C ratio (the activity ratio) yields an age, which may differ significantly from the true age. The differences are nonsystematic, i.e., the activity age may be greater or less than the true age. Moreover, the discrepancies between uncalibrated and true ages generally increase with the true age; differences of ~600 years are possible for material 20,000 years old. The age directly determined from the activity ratio is termed uncalibrated. By comparison with a complicated nonmonotonic oscillatory calibration curve, which is almost annually updated and revised, it may be converted to a calibrated age, quoted as, e.g., CAL BC410. Commonly, these are reexpressed as before present ages. BC410 would be CAL BP 2360 because researchers regard the year AD1950 as the reference point for radiocarbon ages. This is arbitrary but acknowledges the date at which Libby first started radiocarbon studies.

The valid time range over which radiocarbon age may be used is limited by the availability of a technique that permits the activity ratio to be calibrated. In effect, this restricts the maximum precision of ^{14}C to the range overlapping with available dendrochronology or varve chronology, which is approximately ≤15 Ka, at present.

PERIODIC SECULAR MAGNETIC VARIATION (PSV) AND GEOMAGNETIC REVERSALS (GPTS)

The Earth's magnetic field fluctuates in strength and in orientation. Fortunately, due to the science of paleomagnetism, we may measure the ancient orientation of the Earth's magnetic fields trapped in rocks as old as 4 billion years

(4000 Ma) through to materials as young as recent sediments and archaeological monuments.

The outer core is fluidlike and decouples the inner core from the mantle. The turbulence in the geodynamo of the fluid outer core is in some complex way responsible for the fluctuations in the strength and direction of the geomagnetic field. The westwards drift of the geomagnetic field, observed for several hundred years, and recorded archeologically and sedimentologically for thousands of years, is due to mechanical decoupling of the inner core, outer core, and mantle. The consequent difference in rotation rate may explain the westwards drift of the geomagnetic field, the magnetic poles circuiting the geographic poles about once every 1000 a (the Bullard cycle). Continuous PSV records document variations in field direction consistent with westwards drift over the last few tens of thousands of years. Thus, at any given site, the declination and inclination of the geomagnetic field pulse about average values. The periodicity is imperfect so that a curve showing the change of declination or of inclination with time becomes a chronological record. The declination and inclination of the magnetic record in a rock, sediment, or archaeological site are compared with the master curve to determine its age. The measurements are in fact very complicated. Moreover, there is no universal PSV record, "master calibration curves" are valid only over an area of perhaps 500 × 500 km.

Geomagnetic reversals are a complete switch in the direction of the geomagnetic field. They occur when the Earth's magnetic north pole and magnetic south pole switch locations; the bipolar states of varying duration give rise to a geomagnetic polarity time scale (GPTS). The spreading oceans capture this GPTS as patches of normal and reversed polarity on the seafloor. The pattern is approximately symmetrical about the spreading axis of the oceans and may be detected from suitably equipped ships and aircraft (Figures 3.2 and 3.3). The stripes of alternate polarity (Figure 3.2) may be calibrated by geochronology or paleontology and grouped into larger units that date the seafloor (Figure 3.3). Polarity reversals occur synchronously over the entire globe. On average, they appear to have occurred every million years (very approximately) with a few "quiet" periods when the Earth's polarity remained constant for tens of millions of years (e.g., during the Cretaceous quiet period). The alternating magnetic polarity provides a chronological sequence, especially in thick lava sequences (e.g., Iceland), in deep-sea sediment cores and most spectacularly as magnetic stripes all over the ocean floors (Figure 3.2). Seafloor polarity reversals lead to the discovery of ocean floor spreading and to the birth of the plate tectonic paradigm. In brief, as the ocean floor grows and spreads sideways from the mid-ocean ridges, the new igneous rocks cool, trapping the polarity of the ambient geomagnetic field. As increments of ocean floor are added at the ridge (Figure 3.2), they preserved alternating polarities, leading

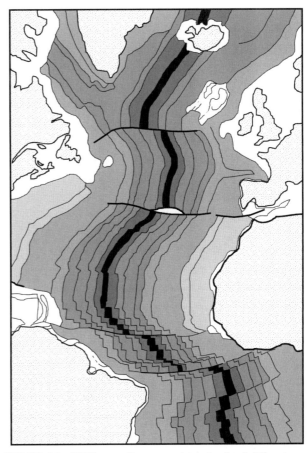

FIGURE 3.3 GPTS anomalies, grouped into bands of different ages across the Atlantic; note symmetry across the mid-Atlantic ridge.

to an approximately vertical magnetic striping (Figure 3.3) of the ocean floor.

GEOCHRONOLOGICAL CONSEQUENCES FOR GEOLOGICAL HISTORY

Using this book, you will be able to learn techniques that may be applied to interpret maps of rocks of whatever age on the Earth. These will reveal the geometry and relative ages of rocks. However, this does not mean that all Earth processes have remained constantly in operation at every stage of Earth history. A principle called uniformitarianism was once rigidly applied (initiated by James Hutton, 1785) in that sense. It is true that the laws of physics and chemistry have universally applied but changing conditions in the Earth mean that geological processes have changed with time. For example, it is doubtful that there was any oceanic crust comparable to that of today in the Archean since sheeted oceanic dike complexes and complete ophiolite sequences are absent. A few important non-Uniformitarian processes that should be borne in mind when studying geology include

1. decreasing heat production, 4 Ga to present (Figure 3.4),
2. life forms increasingly complex and abundant with time,

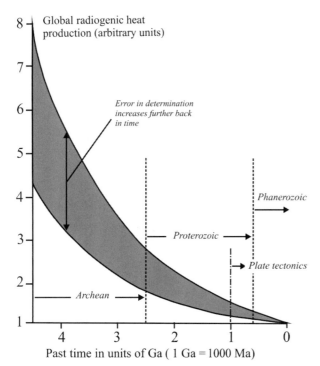

FIGURE 3.4   Heat production due to radiogenic isotopes through geological time.

3. atmospheric changes with time, evolution of the atmosphere and shorter term fluctuations,
4. evolution of the oceans,
5. oceanic sea level changes with time, generally increasing volume over Earth history,
6. oceanic chemistry,
7. nonplate tectonism existed in early Earth history (pre-1000 Ma), and
8. facies series of metamorphism change with time.

Nevertheless, some elementary geology textbooks still promote uniformitarianism as an absolute principle for all processes, not just the laws of physics and chemistry.

Geochronological ages date the closure of the mineral in question to the mobility of a daughter isotope. Certain refractory minerals, such as zircon, retain the Pb daughter product well but many minerals leak the daughter elements. Consequently, they cannot fix the age of the rock, only of some delayed closure event. Examples are very common at the end of regional deformation when regional cooling proceeds slowly producing a geographic dispersion of cooling ages (Figure 3.5). Contours of the ages, chrontours, reveal the uplift pattern and the thermal doming.

Figure 3.6 presents a simple cross-section based on the geology near Thunder Bay, Ontario. By answering the questions, you will see the uses of geochronological data.

1. Using inequalities (> < > < =) bracket the ages of
   a. Shale.
   b. Lava.

c. Deposition of conglomerate.
d. Age of metamorphism.
e. Age of the rocks that were metamorphosed (i.e., protolith).
f. Indicate and date any stratigraphic discontinuity.
g. Describe the nature of the intrusion at (a), (b), (c), and (d).
h. Which eons or eras are present in this era?
i. How would a geochronological age for the lava refine the ages of the other lithologies? If so, what geochronological techniques would be used?

2. Bracket the age range for the following rock types using the inequality and equality symbols (i.e., >, <, ≥, ≤, and =):
   Shale...
   Lava..
   Deposition   of   conglomerate...
   Age   of   metamorphism...
   Age of the rocks that were metamorphosed (i.e., age of the protolith)
   ....

3. Indicate any stratigraphic discontinuity on the diagram and give it the most precise term (i.e., angular unconformity, disconformity, nonconformity).
4. Name and describe the form of the diabase intrusion at
   (a)..
   (b)..
   (c)..
   (d)..
5. Which Eons or Eras are represented by which events and rocks in this area?
   ....
   ....
6. How would a geochronological age for a refractory mineral in the lava assist in determining the absolute ages of the adjacent lithologies?
7. What decay series and mineral would be most suitable in question (i) above, and name one or more decay series that would be less favorable.
8. What would a K–Ar age tell us about the age of the metamorphic rocks? How would this information differ in value from that determined by a U–Pb method?

The absolute ages which bracket the periods are derived from volcanic strata or from igneous units that are truncated by strata. As research has progressed the ages have been revised to smaller values due to the discovery of examples that yield more refined bracketing. For example, in older textbooks (1960s), the base of the Cambrian was believed to be 600 Ma. It is a useful exercise to obtain one or more older texts and fill the empty columns with older estimates.

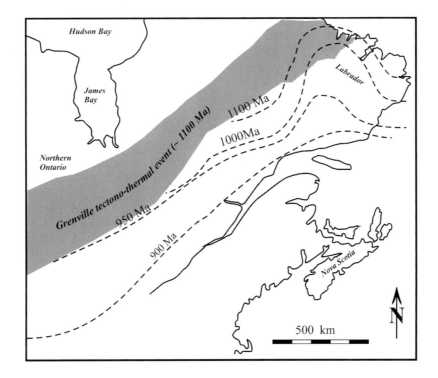

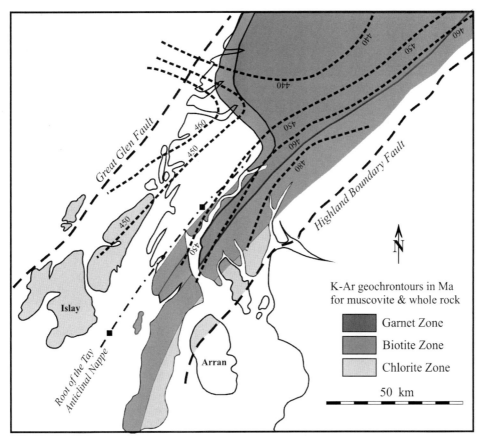

FIGURE 3.5 In metamorphic rocks, as shown here, geochronological methods do not always reveal ages of the original rock (=protolith). Instead, the ages mark the closure of the rock to expulsion of the daughter element. Thus, the ages mark temperatures associated with postmetamorphic uplift and cooling. Lines of equal geochronological age are called chrontours. (a) chrontours for the eastern Candian Shield (b) chrontours for SW Scotland.

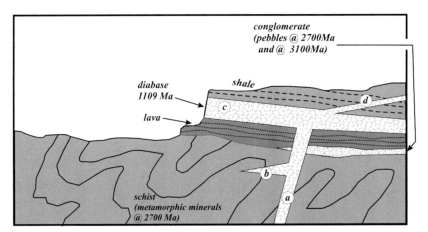

FIGURE 3.6 Examine the cross-section of simplified geology of northern Ontario and answer the questions.

Chapter 4

Age Relationships from Map View

The cross-sections and block diagrams shown previously lend themselves readily to understanding relative age relationships because the rock units and structures are viewed in a vertical cross-section. Most of the time, geologists infer age relationships, i.e., relative ages, by inspection of the map view alone. The cross-sections such as those shown in previous diagrams are constructed only after the geologist has studied the area.

The pivotal criterion used to infer relative ages is the crosscutting relationship. It is the one criterion that may be potentially applied to every map. Its usefulness depends on the fact that the crosscutting relation viewed in a vertical cross-section must is almost always visible as a crosscutting relationship at the surface. The "T" form of the crosscutting relation may be skewed in map view, just as it is in cross-section, depending on the inclination of the planes with respect to the surface on which the junction is viewed.

Figure 4.1(a) illustrates an angular unconformity in cross-section at the front of the block diagram; the shale overlies the other beds forming a "T"-junction. It is clear that the map view must also show some aspect of the T-junction, as shown on the top surface of the block in Figure 4.1. Note that the geologist only records orientations of beds with respect to the horizontal surface and records them only on that horizontal (map) projection. Note also that an angular unconformity may occur if the beds have the same dip above and below the unconformity; a difference in strike also produces an unconformity. Note that the T-junction may not be present in every projection. For example, in the map view of Figure 4.1(b), it cannot be seen from the map alone.

The T-junction concept may be applied to depositional contacts between sedimentary rock units but also to the contacts between intrusions and other rocks. An igneous intrusion (as in Figure 4.1(a)), a forceful intrusion of rock salt or a forcefully emplaced plutonic batholith will usually abruptly truncate earlier layers and contacts in the country rock. The T-junction principle will only fail to help us where the layers on either side of the contact have a similar strike (Figure 4.1(b)).

An exception to the application of the T-junction principle may occur with fractures such as faults and joints. Generally, a modestly sized fracture will act as a free surface across which later fractures cannot propagate. Consequently, younger fractures, especially smaller faults and most joints, terminate against older fractures (Figure 4.1(c)).

Without any independent evidence of stratigraphic order, i.e., no stratigraphic column or ages, it is still possible to infer the relative ages from most maps. Figure 4.2 shows two excerpts from the Thunder Bay Quaternary Geology Map of northern Ontario, a map that emphasizes the glacial and other superficial deposits that overlie the Precambrian bedrock. From these two maps, using "T"-junctions and perhaps using the simplified terminology below, infer a relative age sequence for these glacial sediments.

Lacustrine = deposited in a lake, e.g., at the margins of a glacier, includes mud and varves.

Moraine = poorly sorted material (boulders, sand, and clay) melted out from a glacier.

End moraine = (terminal moraine) melt deposits at terminus of more-or-less stationary glacier.

Ground moraine = deposited or reworked beneath glacial ice.

Outwash sand = channel deposit caused by melt water from glacier.

Loess = windblown fine material, eroded from glacier by spring windstorms.

Varve = cyclical annual fine sediment washed out from glacier into marginal lake.

Figure 4.3 shows a hypothetical map with stratigraphic, igneous, and tectonic contacts. The area is large and topographic relief is negligible. Identify the following (there may be more than one of each): angular unconformity, disconformity, nonconformity, onlapping sequence, passive intrusive contact, forceful intrusive contact, tectonic contact, xenoliths, dike, and sill.

Understanding Geology Through Maps. http://dx.doi.org/10.1016/B978-0-12-800866-9.00004-1

(a) T-junction with rock units

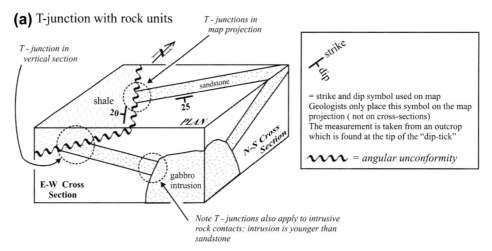

(b) No map view of T-junction if strikes same

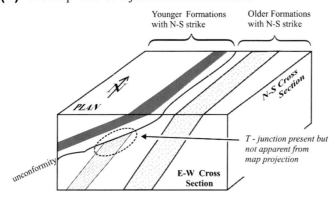

(c) T-junction with faults (or joints)

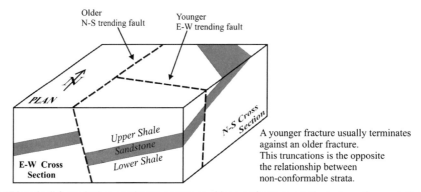

FIGURE 4.1 The "T-junction" formed where one layer truncates an older one (fractures may be an exception see (c)). (a) T-junctions in plan and cross-section. (b) T-junction not apparent at surface.

The heavy black lines mark diabase sills/dikes. Remember that a sill is concordant with strata and a dike is discordant; the orientation of the sheet is irrelevant although, in general, sills are gently sloping.

Figure 4.4 is a simplified map of Thunder Bay Quaternary glacial deposits. Figure 4.2 gave two tiny excerpts from the region shown here. This map is more informative.

How many terminal moraines are there in this area?
Show the directions of advance of the glacier(s).

Which way did the glacial ice retreat?
Which way did the glacier advance?
What is the most common type of glacial deposit?
Which glacial sediment has probably been least well preserved?

Figure 4.5 shows the simplified submarine geology of superficial deposits in Lundy Bay off the coast of South Wales. The land geology is not shown and the land area is left entirely unornamented. Milford Haven and in the SE,

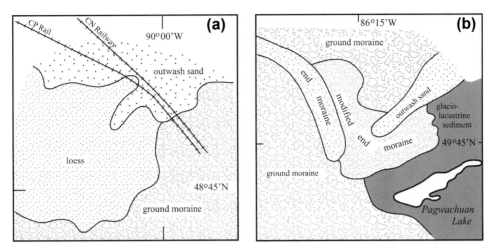

FIGURE 4.2 Two excerpts from the Thunder Bay (NW Ontario) glacial map showing the distribution of glacial deposits. T-junctions indicate the relative ages of the sediments. Answer the questions in the text for maps (a) and (b).

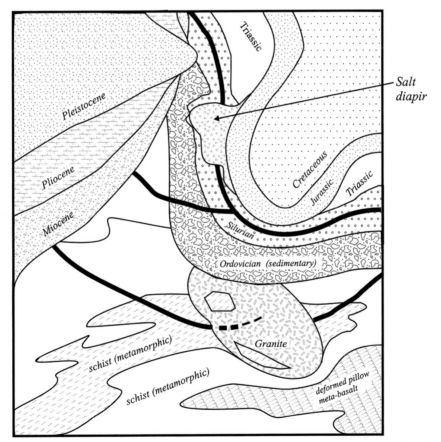

FIGURE 4.3 Hypothetical map showing stratigraphic tectonic and intrusive contacts. Answer the questions in the text.

the Bristol Channel (fed by the River Avon), provide massive stream input. The Irish Sea is to the west. All the materials mapped on the seafloor are unconsolidated sediments, with the exception of small gray areas where bedrock has been exposed by submarine erosion. Some sediment types overlap or overlie upon other sediment.

The map was made by dredging sediment samples and by coring the sediment column from ships which has great economic benefit since there is much offshore petroleum around Britain. The legend is plotted on a triangular diagram that shows the proportions of sand mud and gravel in each sediment type. Ask your instructor to explain the

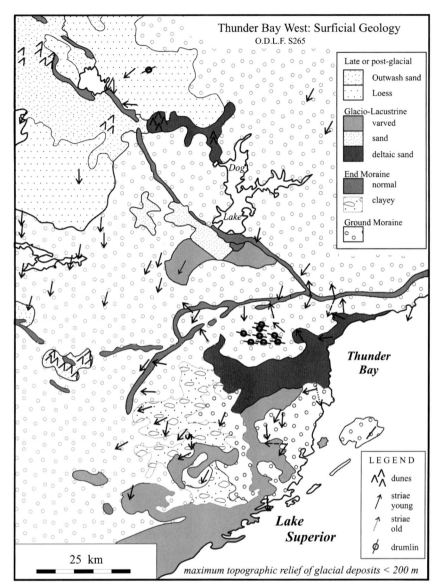

FIGURE 4.4 Glacial deposits near Thunder Bay, NW Ontario (Map by Zoltai, 1965). Using T-junctions, and the nature of the sediment, answer the questions in the text.

interpretation of a triangular diagram; it is very important in many aspects of geology since it expresses the content of a material in terms of the proportions of three components.

From the legend, note also the explanation of the average current–velocity symbol. Such information is quite rare and difficult to obtain but assists in the interpretation of the distribution of types of sediment. (In general, higher current velocities move larger grain-size sediment.)

1. Are the sediments clastic, chemical, or organic? (Consider legend.)
2. Approximately 10 km offshore from Milford Haven, sand deposits in the SE abut bedrock, to the NW of which is found gravel. Can this be related to current velocities in the area and which is the dominant direction of current flow?

3. Can you explain why mud is deposited in the bay between Ramsey Island and Skomer Island? What is the stratigraphic order of the sediment in this bay? (Sketch a simple column, oldest at base, as always.)
4. Indicate one area on the map where bathymetry (submarine topography) appears to control the pattern of sedimentation. Indicate another area in which the sedimentation is quite unrelated to bathymetry. Why is that so, in the latter case?

Figure 4.6 shows the very simplified geology near Attica, Greece. Since the map is of a rather large area, we can disregard topographic effects and make a reasonable geological interpretation from plan view, making the simplifying assumption that the view is of a horizontal plane. Geological contacts vary from steep to nearly horizontal and the

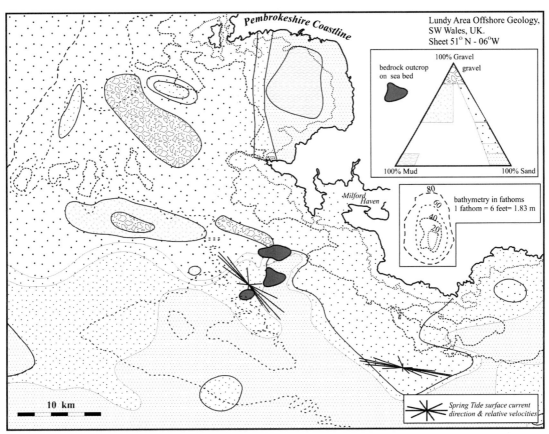

FIGURE 4.5 Submarine deposits off the coast of South Wales. Using T-junctions, and the nature of the sediment, answer the questions in the text.

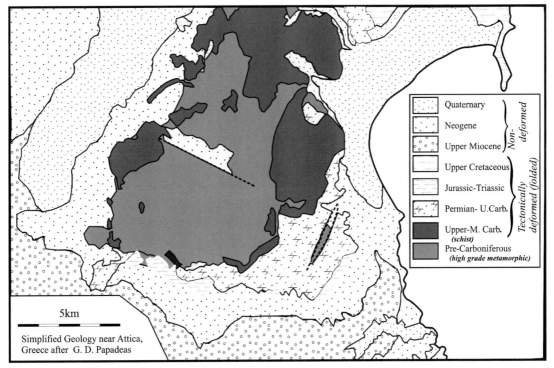

FIGURE 4.6 Geological map of Attica, Greece.

dendritic pattern of the youngest horizontal deposits may be used to identify them. The high-grade metamorphic rocks and schists are generally "surrounded" by younger rocks, from which we infer that the geological sequence is arched upward in a dome located at the center of the area.

Q.1. From the truncation of boundaries ("T-junctions") or from lithological considerations identify four stratigraphic levels at which unconformities (or nonconformities) occur. Indicate these on the stratigraphic column. Indicate locations of critical T-junctions on the map.

Q.2. What features influence the outcrop distribution of the Quaternary and Neogene sediments.

Q.3. The nondeformed strata were eroded from the older rocks as they were "domed". What evidence is there that this uplift (doming) was incremental rather than having occurred in a single episode?

Q.4. Make a sketch cross-section west-to-east with west on the left, showing the approximate arrangement of the beds some distance underground. You may consider all dip angles to be modest but the relative dip angles and directions can be inferred from the map.

In Figure 4.7, you may construct a simple geological map of a horizontal plane, using the information provided in the cross-sections. Note that the slope of any given bed is different in each section because the sections are not parallel to the dip directions. In other words, apparent dip directions are shown in the cross-sections. This is also the reason that the same beds appear to have different thicknesses in each section.

Figure 4.8 requires to complete the map using the information from the two cross-sections. This is also has zero topographic relief. A simple beginning is made by joining the base of the conglomerate on the east and south side off the map. Complete the map of beds above the unconformity first. With simple reason, the other bedding planes may be completed across the map. Since the dike does not appear in the N–S section you are free to chooses from a wide range of valid paths to draw it across the map. Why do the same strata show different dips in the two sections?

Figure 4.9 requires the completion of a map that introduces the complication of topography. Since horizontal beds show complete sympathy with topographic contours you may take the base of the shale (400 m) and map it across the map following the 400 m topographic contour precisely. Similarly, you may trace the base of the limestone across the map following the 200 m contour. Mapping the diabase dike and sill requires a little more thought. Since the dike is vertical, it shows no sympathy with topography and cuts a straight line across the map.

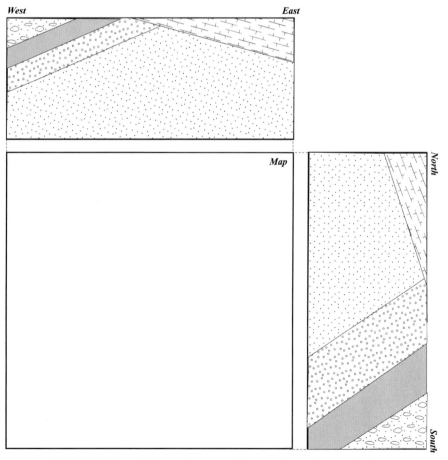

FIGURE 4.7   From the two sections, complete the geology of the map. See text.

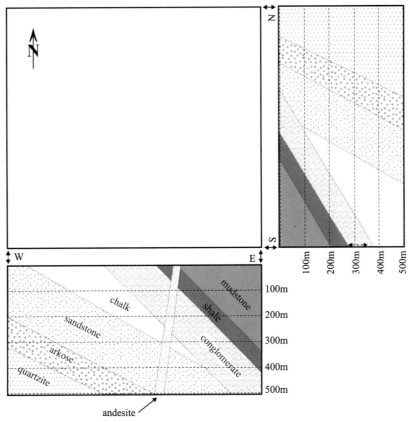

FIGURE 4.8   From the two sections, complete the geology of the map. See text.

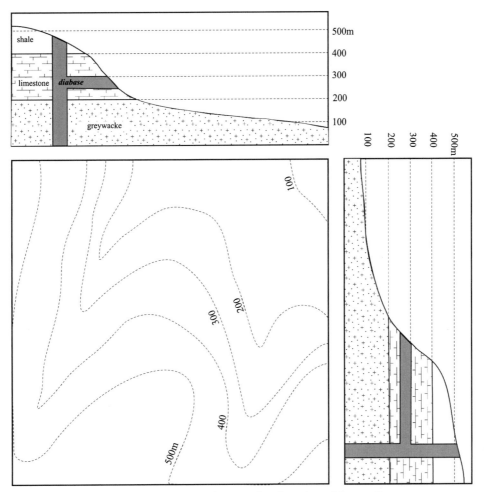

FIGURE 4.9   From the two sections, complete the geology of the map. See text.

Chapter 5

Layered (Stratified) Rocks and Topography

Chapter Outline

In the previous maps, we saw that without examining the topography of the map, it was possible to infer relative ages of rock units. Since topography was not considered, it follows that those principles apply almost regardless of the scale of the map. However, most Earth scientists work on scales on which the maps show topographic relief and the variation in elevation provides more information on the rock units because they are exposed at different locations according to their elevation that, in turn, is controlled by the dip of their contacts. Now, we shall examine maps on which topographic relief is important in controlling the shape of the rock units on the map. Crosscutting relationships and the "T-junction" principle are still valid for inferring relative ages from the map view of topographic slopes but the angular discordance will appear smaller on the map than if the ground were horizontal.

The approach initially is to use simple geometrical constructions that compare topographic contours with hypothetical contours that reveal the slope of strata; the latter are called stratum contours or structure contours. However, experienced geologists are able to "read" the geological map, comparing topography and the sinuous nature of geological boundaries to visualize the dips of rock formations and the relative order without any further paperwork. It should be noted that many published geological maps by government geological departments have all the geological information superimposed upon existing topographic maps. Thus, it requires considerable concentration and good eyesight to interpret them; they are intended for study, not for browsing general classroom use or use in the field.

Figure 5.1 reveals the essence of the above discussion. The diagram shows a small valley and a stratum mapped in and exposed on the cross-sections of the block. In (a) the stratum is horizontal and thus its top and base perfectly follow topographic contours, the outcrop V's into the hillside. In (b), with the same topography, a vertical bed shows no sympathy with topography and cuts straight across the map, without deflections due to topography. Dipping strata only show minor complications; in (c), a bed dipping into the hill will show an outcrop that V's upstream and it will transgress the topographic contours gently in the fashion shown. In (d), a stratum dips downstream, more steeply than the slope of the stream, and as a consequence it V's downstream. The different sense with which the bed crosses the contours in (b)–(d) should be remembered and the best way to do that is to learn to sketch them for yourself.

Of course, without some geometrical construction, this principle does not inform the reader about the angle of dip of the bed; from the map view with a careful consideration of scale, it may be possible to rapidly estimate if the dip is >45° or <45°. Therefore, published geological maps also carry symbols that indicate the inclination (dip) and strike of beds. Strike is the trend of a horizontal line drawn on the bedding plane. The dip angle occurs at the tip of a tick on the dip-and-strike symbol shown at the base of Figure 5.2. Note the special symbols for vertical and horizontal strata. Geologists record the azimuth of the strike as well as the dip angle but maps mostly show only the dip angle. The dip angle ranges from 0 to 90°, whereas strike, being an azimuth, may range from 0 to 360°. The tip of the dip symbol identifies the location for which the dip and strike are valid. This is important to remember because the dip-and-strike symbol make cover several hectares of ground on the map.

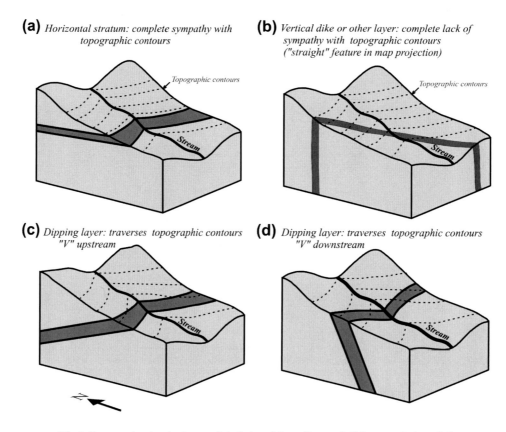

(a) *Horizontal stratum: complete sympathy with topographic contours*

(b) *Vertical dike or other layer: complete lack of sympathy with topographic contours ("straight" feature in map projection)*

(c) *Dipping layer: traverses topographic contours "V" upstream*

(d) *Dipping layer: traverses topographic contours "V" downstream*

Block diagrams showing the degree of similarity of shape ("sympathy") between the boundaries of outcrops and the topographic contours. From a comparison of the outcrop boundaries and topographic contours, it is therefore possible to infer the direction and approximate amount of dip of geological boundaries. This applies to all manner of geological boundaries, including faults, and intrusive contacts.

If you were to view each of the above situations in map projection, which of the following strike and dip symbols would be best associated with which diagram?

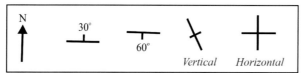

FIGURE 5.1   Topographic expression of a stratum, depending on its dip and the dip of the valley.

Figure 5.2 shows a large area of the Grand Canyon. This large area map has extreme topographic relief but relatively simple geology with steep contacts in the Archean rocks and almost horizontal strata in the Proterozoic and Phanerozoic rocks. The outcrop pattern of the Cambrian and Mississippian rocks is characteristic of horizontal strata since they follow the stream courses and topographic contours in a dendritic fashion. Draw two cross-sections with an exaggerated vertical scale, one N–S through the center of the map and another E–W through the center of the map. Locate the unconformities/nonconformities by inspecting the legend and searching for T-junctions.

You now understand enough of the basic principle of the interference of topography with geological planes you may attempt the problems sketched in Figure 5.3. The upper map shows three differently dipping diabase sheets and a partially mapped sandstone bed. The sandstone dips 20° to the NNW. Which of the diabase sheets is a sill, and which are dikes? Estimate the path of the sheets across the map from the degree of their sympathy with topography. Mark the top and the base bedding planes of the sandstone. Extrapolate the outcrop of the sandstone using your knowledge of its sympathy with topography.

Note: a representative fraction 1:10,000 gives the scale, i.e., 1 cm on the map represents 10,000 cm (100 m) in the field.

The lower map in Figure 5.3 shows topographic relief, two diabase sheets, and a fault. What can you infer about the orientation of the two diabase sheets? Indicate the

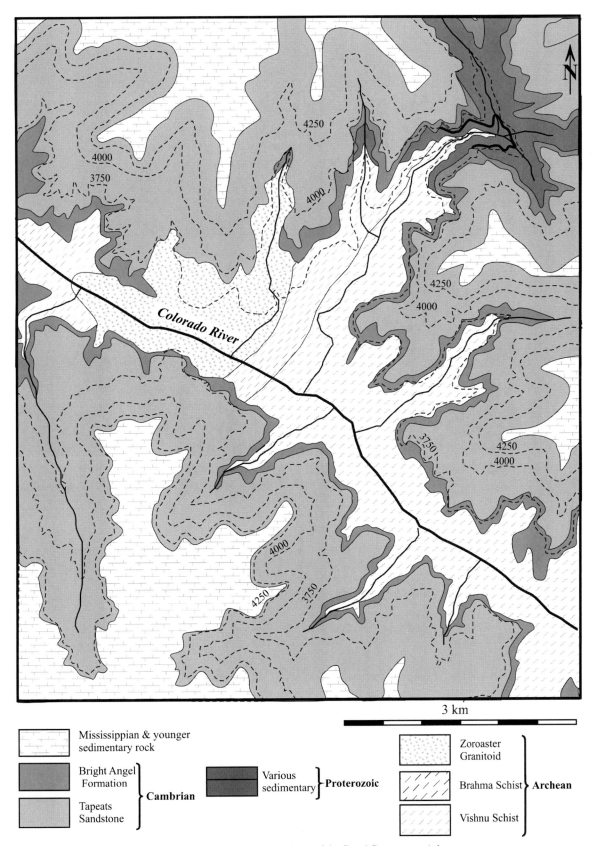

FIGURE 5.2 Simplified geological map of the Grand Canyon area, Arizona.

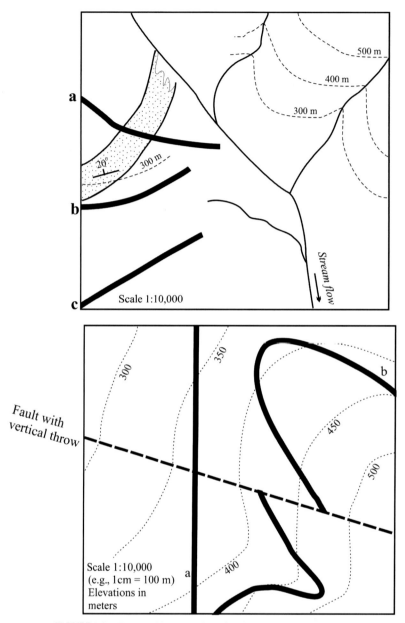

FIGURE 5.3   Topographic expression of various lithological features.

dip and strike of the sheets using the traditional symbols. Are the intrusions older or younger than the fault? ("a" is ambiguous). Construct structure contours (ask your instructor) for sheet "b". What can you say about its dip and strike?

DIP, STRIKE, AND THEIR MAP REPRESENTATION

The relationship between dip, strike, and dip direction is simplified in Figure 5.4(a) and the map representation is shown in the adjacent sketch (Figure 5.4(b)). The preferred dip-and-strike symbol requires some modification

for vertical and horizontal beds as shown. Its modification in this way shows how it is superior the older "dip-arrow" map symbol, which cannot be modified in consistent manner to represent vertical or horizontal beds. There is little confusion in reporting vertical orientations (e.g., 045/90SE and 045/90NW are equivalent). However, for horizontal beds, a notebook record should simply state "horizontal"; strike is indeterminate and one is free to choose any values from 000 to 360°.

For future reference, the serious student may be interested to note that orientations may be brought together regardless of their location and compared in an orientation diagram known as a stereogram (Figure 5.4(c and d); also

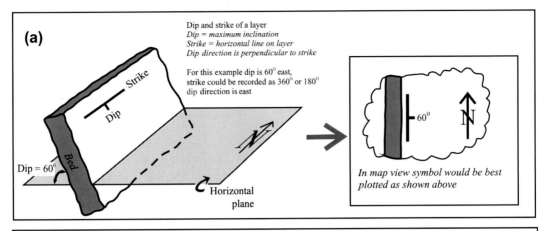

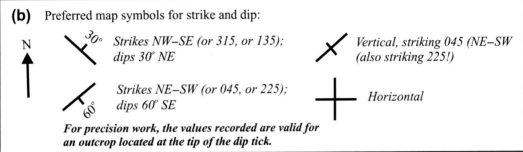

For future reference (also see appendix):
The orientations of beds and other planar structures may be taken from maps and used for advanced geometrical or statistical studies in structural geology.
This uses a hemispherical projection (*sterographic projection*) of the orientation of the plane. The plane is recorded as if contained in a hemispherical bowl that is viewed from above.

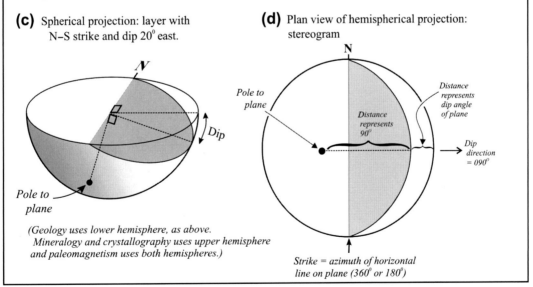

FIGURE 5.4 Orientation of strata. (a) definition of dip and strike (b) map symbols (c) for advanced work (structural geology) a stereogram projection may be used.

see Appendix). This is a hemispherical projection viewed from above. The orientation of the plane (or line) is located at the top center of the hemisphere and projected down to the spherical surface. When viewed in plan, the circular projection of the hemisphere, known as a stereogram or stereographic projection provides a powerful tool for solving structural problems and statistically evaluating field data, microscope data of crystal orientations (="petrofabrics") and geophysical information (e.g., paleomagnetic directions).

RECORDING ORIENTATIONS: CONVENTIONS

Primary geological structures that constitute a planar orientation include beds, strata in general igneous dikes and sills (Figure 5.5). "Strata" encompass layered igneous rocks such as lava flows and certain internally layered igneous intrusions. "Primary" refers to features formed by the initial geological process, prior to any modification by tectonic deformation or compaction. For example, all beds commence as planar features, at least approximately, but secondary events may fold them in mountain belts or in slumped sediments.

Some secondary structures are also planar features; faults, tension joints, and most shear joints. In metamorphic rocks, cleavage and schistosity are planar structures caused by secondary alignments of new minerals. The central part of ductile shear zones may also be quite planar, usually being schistose.

Geometrically considered, a plane is a three-dimensional feature and three pieces of information are always required for its definition (recall its mathematical definition, $\mathbf{a}x + \mathbf{b}y + \mathbf{c}z = 0$) (Figure 5.5(a)). However, as geologists, we have adopted conventions for recording planar orientations that reduce the information to just two numbers. This has become so ingrained in our training that we forget that the third piece of information is, in some systems of reporting,

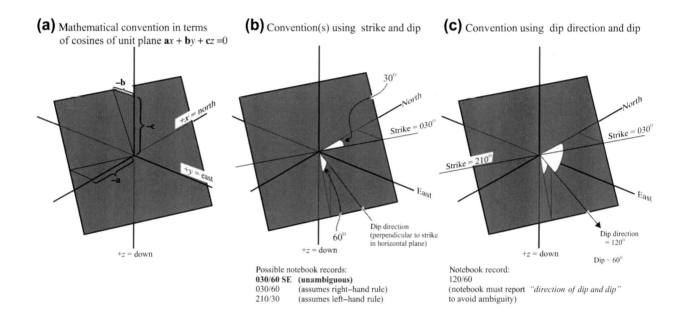

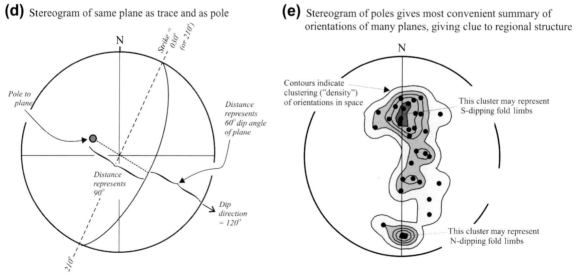

FIGURE 5.5   Advanced terminology for orientation of strata.

disguised and only apparent to other geologists who are familiar with the system.

Fundamentally, the plane is located in a coordinate system where $+x$, $+y$, and $+z$ represent the directions north, east, and "down". Geophysicists, from necessity commonly need to use the coefficients (**a**, **b**, **c**) to define the orientation of a plane but this convention would not be very convenient for field records of planes whose orientations are measured directly in angles using a compass. (Note: the signs of x, y, and z are important and the plane is a unit plane which only records orientation, i.e., $x^2 + y^2 + z^2 = 1$).

The geologist records the plan's orientation in one of two ways. The most common records the strike and dip (Figure 5.5(b)). The strike is the direction of a horizontal line on the plane. Its azimuth is recorded from the compass bearing and, being an axis rather than a direction, it is equally valid to report the example as having a strike of 030° or 210° (i.e., 30° + 180°). The dip of the plane is the angle of maximum slope of the plane, measured from the horizontal, in the example shown it is 60°. The least ambiguous way to report this in a notebook is to write

030/60SE

where "SE" removes the ambiguity as to which way the bed dips relative to the strike line; without adding "SE", it would not be known if the bed actually dipped to the NW.

However, many geologists unwisely prefer a convention in which the three pieces of information (030, 60, and "SE") are embedded in a coded form which can be reported with just two numbers. For example, most North American geologists use a right-hand rule in which this plane is reported as 030/60. This system assumes that the dip direction is always found in an anticlockwise direction from the end of the strike line (030) which is reported. Moreover, it is not universal, for example, some Australian geologists use a left-hand rule in which the same bed's orientation would be reported as 210/60. These shorthand "rules" are subject to error, not universally understood and most of their supporters do not even write in their notebooks which convention was used.

As we shall see later, both of these unwise conventions introduce further confusion when the geologist has to report the orientation of a lineation or linear feature. This is successfully and uniquely specified with two numbers such as 180/25 defining the trend (180° or South) and plunge (25°) of the lineation (Figure 5.8).

A third manner of recording the orientation of a plane relies on measuring the dip direction and the dip (Figure 5.5(c)). This method is actually more convenient for those using a stratum compass. The stratum compass reduces labor and increases precision as we shall see below. However, it is more useful to show the strike direction on a map than the dip direction, using the symbols recommended because the trend of the planes is immediately apparent to the map reader.

Note that, to avoid confusion, azimuths must always use three digits and dips must always use two digits to reduce confusion. For example, if the strike was only 6° east of north, and the dip angle was 4° to the east, it would be rather puzzling to read "6/4E" in a notebook. The report "006/04E" would be less ambiguous.

HOW DOES THE FIELD GEOLOGIST MEASURE THE ORIENTATION OF A PLANE?

The dip-and-strike symbols reported on published maps are largely determined in the field by observation and measurement with a specially designed magnetic compass (Figure 5.6). The "geological compass" takes several forms and costs several hundred dollars at current prices. In order of decreasing frequency, one finds models that encompass the following features above the basic requirement of a compass needle that points to North.

1. All contain a scale, compass needle, and a separate non-magnetic free-swinging needle that permits dip angles to be determined; this is called an inclinometer or clinometer.
2. One or more bubble gauges may be fitted to ensure that the dip angle is indeed measured with respect to the horizontal.
3. A lid or edge that may be placed along the strike of a plane, to facilitate in measuring the strike's azimuth. An appropriately placed bubble gauge assists here.
4. The lid may itself be an inclinometer. Thus, the lid may be placed on a sloping surface and the dip and direction of dip of the surface may be measured in one action. (Most compasses require strike and dip to be measured in two separate actions, and most geologists prefer to record orientations as strike and dip rather than dip and direction of dip.) These compasses are called stratum compasses (Figure 5.6(a)).

Furthermore, some of the following features related to the compass needle and its reading may be present.

5. An adjustment for magnetic declination of the compass needle (the azimuth of true or geographic north differs from magnetic north almost everywhere).
6. A compass scale that reads anticlockwise from north. This is shown by the azimuth numbers but also most disturbingly for the beginner by the exchange of the symbols "E" and "W" for east and west! As the compass is turned to make its edge parallel to some significant geological direction, the compass needle then becomes direct reading, pointing to the azimuth number appropriate for the geological direction.
7. Some device that permits the compass to be used for "sighting" or "reading a bearing" to some prominent topographic feature such as a cairn. This may take the

(a) Stratum compass (German and Japanese)

This is generally considered to be superior for making routine structural measurements since dip and direction of dip are recorded in one manipulation of the compass. The slope of the lid records the dip of a plane (or plunge of a lineation) and the trend of the compass edge gives the direction of dip (or trend of a lineation). A graduated protractor on the lid hinge gives the dip and color codes it to the correct direction of dip. Strike may be recorded by mental arithmetic from that one-step measurement or by placing the compass edge along the strike line of the plane. Usually one or two level guages permit accurate leveling; the Japanese version also permits compass sighting to be made (for map positioning) as with the transit (Brunton style) compass.

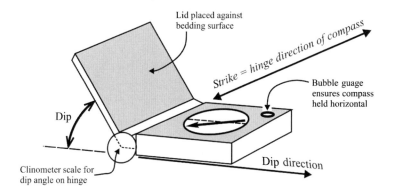

(b and c) Transit (Brunton style compass)

Modified for geological use from the traditional surveyor's compass, this was primarily designed for compass sightings so that *map positions* could be determined by triangulation. That purpose has now been superceded by GPS. (b) Strike is measured using the compass edge and (c) dip is measured using a clinometer which swings to the vertical on the backplate of the compass. Again, level gauges ensure horizontality in (b) and verticality in (c).

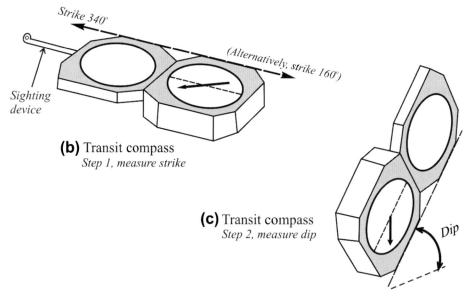

FIGURE 5.6   Use of the geological compass.

form of a device like a "gun sight" which folds out from the compass or a hole in the lid of the compass; a mirror may be used to simultaneously sight on the object and read the compass needle bearing (analogous to the structure of a sextant). Compasses which show this feature are sometimes called "transit style" compasses since they are modified from land surveyor's compasses. They also often take the name of the most prominent manufacturer, "Brunton style" (see Figure 5.6(b and c)).

8. An adjustment to ensure the compass needle remains horizontal at different latitudes. (Northern hemisphere geologists who travel for work in the southern hemisphere find that the needle of their compass points upward so steeply that it is trapped by the glass dial and will not rotate if the compass is held in its correct horizontal orientation!)

9. The presence of an eddy current damping conductor around the edge of the compass ring suppresses unnecessary oscillations as the compass comes to rest at its north-pointing position. Traditionally, as in old marine compasses, this damping was achieved with low viscosity oil within the compass housing. However, it is not favored in geologists' compasses due to the added weight and risk of loss during the rugged handling required of a geologist's compass.

Digital electronic enthusiasts, please note that global positioning system (GPS) units do not provide a reliable means of measuring orientations of field structures! Also, attempts to market magnetic compasses with digital display have failed because the underlying mechanism relies on mechanical and magnetic interactions anyway. Such digital compasses give a false sense of accuracy, are dependent on batteries, and require careful construction so that electrical currents do not interfere with the operation of the magnetic needle. In comparison, the traditional analog compass needle leaves one with no false impressions of precision; directions are rarely measured to better than ±3° with a handheld compass and dip angles may be measured no better with its inclinometer. Indeed, most geological surfaces are so rough that it is not reasonable to expect better precision and commonly precision is much worse than ±3°. For many rough surfaces, it may be wise to lay a clipboard on the surface, which acts to average out its roughness, and then measure the orientation of the clipboard.

Of course, to locate position on a map with a small handheld device, the use of the magnetic compass has been superseded. An inexpensive GPS locates position at least within meters. In contrast, the geologists' small handheld transit compass to take bearings on topographic features and triangulating to fix one's position within a triangle of error was often inaccurate and rarely had a precision better than ±5 m when mapping at a 1:10,000 scale. Rarely, of course, GPS units do not work. For example, in areas of severe topographic relief, where the landforms obscure the line of sight to one of the Earth-orbiting GPS satellites. This unavailability may depend upon time of day due to satellite positions above the Earth. Finally, near some military installations, GPS service fails altogether for security reasons.

The beginner to cartography and geology should not worry about the accuracy of published topographic base maps. Features are just as precisely located upon them as if the best GPS system was used. Geologists routinely use topographic base maps made in the 1800s, with confidence. The original map surveyors located topographic features with precisions of centimeters using massive precision scientific instruments called theodolites; they determined positions relative to other topographic features. Latitude was determined precisely with respects to stars using large sextants and longitude was fixed with precise chronometers. Land surveyors of the prescientific age used instruments much more substantial than the equivalent models used on ships. The largest theodolites used by the British Ordnance Survey in the nineteenth century used theodolites weighing in at several tons and required a team of horses for transport!

On a final note, we should be aware that for the determination of elevation, the small handheld GPS unit still cannot always compete with the aneroid barometer. The aneroid is a disc-like chamber made of thin metal from which all air has been evacuated. As air pressure changes, so the surface of the disc flexes like the surface of a drum. The small motion caused by pressure changes is easily amplified by levers to read on a scale. The aneroid barometer was invented in 1843, and, like mercury tube barometers, it was designed primarily to record atmospheric pressure. However, atmospheric pressure varies due to two factors, obviously weather but less obviously altitude. If the atmospheric conditions are constant, the barometer responds to lower pressures as one carries the barometer ascends to greater altitudes. Even small, handheld barometers accurately detect elevation changes of <1 m. (Hold one in your hand and walk upstairs!) The aneroid must be read at some base station of known altitude and then the change in reading will permit one to determine the altitude of an outcrop. Unfortunately, in some climates and in certain seasons, atmospheric pressure may change significantly during the course of the day so that the aneroid should be used to determine relative elevations from frequently checked "base stations". (I took a lunch break on a Spanish mountain and watched the aneroid barometer apparently report that I had climbed 10 m while sitting.)

THE "WAY UP" OR POLARITY OF STRATA: YOUNGING

If you pursue this book far enough, and certainly, if you follow more advanced geology courses, you will appreciate that many strata contain internal features which indicate the younging direction or which is the top of the stratum (Figure 5.7). This feature cannot be identified from a map, as it requires careful observation of some primary depositional feature in the outcrop. Of course, it may be shown on the map and that is the purpose of this paragraph and Figure 5.7. We should be careful to note that some authors incorrectly substitute the term "facing direction" for "younging direction". Historical

(a) Symbols for orientation of bedding or stratification

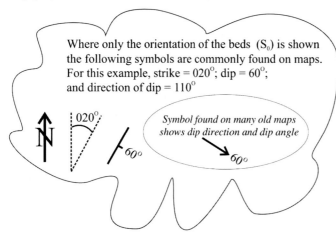

Where only the orientation of the beds (S_0) is shown the following symbols are commonly found on maps. For this example, strike = 020°; dip = 60°; and direction of dip = 110°

020°;

Symbol found on many old maps shows dip direction and dip angle

60°

60°

Note that the symbol for bedding orientation or any other structure occupies a large map area. For example, on a 1:50,000 map (1 cm = 500 m), a dip and-strike symbol may occupy an area of 200 m × 200 m. The map location at which the measurement was made is usually found at the tip of the symbol.

E.g., actual outcrop was here

The following is for future reference and is not essential reading at this point.

(b) Symbols for orientation and polarity of bedding

It is important for field geologists to note the top and bottom of beds in the field from sedimentary and other primary internal strictures of stratified rock.
Some "way up" structures that show the direction toward the top are shown in this cross-section cartoon:

Modern maps place the "Y" symbol with its "leg" pointing the younger beds.

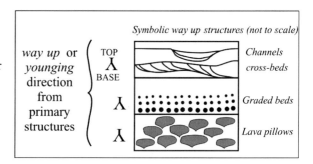

Symbolic way up structures (not to scale)

way up or *younging* direction from primary structures

TOP
Y
BASE

Channels

cross-beds

Graded beds

Lava pillows

The following sketches show map symbols for a bed striking 030° and dipping 35° SE

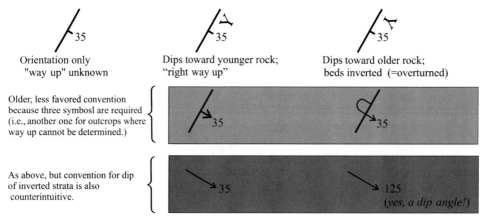

35   35   35

Orientation only "way up" unknown

Dips toward younger rock; "right way up"

Dips toward older rock; beds inverted (=overturned)

Older, less favored convention because three symbosl are required (i.e., another one for outcrops where way up cannot be determined.)

35   35

As above, but convention for dip of inverted strata is also counterintuitive.

35   125
(yes, a dip angle!)

FIGURE 5.7   Map symbols for strata.

precedence shows that "facing direction" or more fully "structural-facing direction" is a more complicated and very powerful observation of great use in folded and complicated terranes; its discussion is postponed to a structural geology course.

The top and bottom of sedimentary layers are readily identified by sedimentary structures such as graded beds (most turbidity currents produce coarse-based beds and finer grain size at the top). Other structures include erosional markings, trough cross-bedding, asymmetric ripple marks, mud cracks, syneresis cracks, sand dikes, load casts, certain burrows, and grazing trails of submarine organisms and rain prints. Some of these structures are identified in cross-section (e.g., see Figure 5.7(b)), whereas others are confined to the top or basal bedding plane of a particular bed.

Lava flows also show some younging structures. For example, the tops of flows may show characteristic ropy or aa–aa surfaces, the bases of flows may show pipe amygdales, and vesicles and gas pipes within lava flows may only partially fill their basal portions with secondary minerals like quartz, calcite, and chlorite. Lava pillow shapes give the most common way up structure in the entire World. Submarine basalts cover at least 70% of the Earth's (submarine) surface, extruded as blobs into seawater which immediately gave them a temporary viscoplastic coating. Since the interior of the pillows (="blobs") is viscous, the pillows roll and settle into the troughs between older pillows, giving them a characteristic V-shaped root, which points downward (Figure 5.7(b)).

The younging direction is easily reported on a map but it is an independent observation that is best not merged with a bedding symbol. Unfortunately, may traditional schemes use a dip-and-strike symbol for bedding and another one where the way up is also observed. This is further complicated where, as in many folded terranes, the bed is overturned. On the other hand, if we keep one symbol for bedding orientation (the dip-and-strike symbol) and one for younging ("Y"), life is much simpler. Examination of the lower part of Figure 5.7(b) should clarify this.

TECTONIC PLANAR AND LINEAR STRUCTURES

For completeness and future reference in structural geology, some typical symbols are shown for tectonic structures (Figure 5.8). There is no single widely accepted convention for representing these structural orientations on maps but the symbols shown are intuitive and fairly commonly used examples. Planar features designated S_1, S_2, etc. are schistosties or cleavages or gneissic foliations. In low-grade metamorphic rock, S_1 is typically a slaty cleavage and S_2 is a crenulation cleavage. In mylonite, commonly the schisosity (S) is interleaved with shear bands (C) to define a composite and more or less synchronous S–C fabric.

Linear elements (L) are of two distinct types. The first is a mineral or shape lineation that is an integral part of a common metamorphic schistosity; they are apparent at all metamorphic grades. The combined fabric is known as an L–S fabric and if the linear component dominates, it is described as L > S; alternatively a strong S and weaker L is described as L < S. Fabric lineations commonly define finite strain direction (X in the XY plane, where $X \geq Y \geq Z$).

The second type of lineation occurs where two planar structures intersect (Figure 5.8(b)). These are most common in low metamorphic grades. For example, bedding (S_0) intersects with slaty cleavage (S_1) to define a first generation lineation l_1. Such intersection lineations may be observed on either of the two intersecting planes. Intersection lineations commonly define the plunge direction of a fold to which the S_1 surface is axial planar.

ORIENTING SPECIMENS RETRIEVED FROM THE FIELD

For many purposes, in structural geology, metamorphism, and paleomagnetism, the geologist must retrieve a sample of rock, preserving its orientation in geographic coordinates (Figure 5.9). That is to say, it must be possible to hold the specimen in the laboratory and know which direction is north and what plane is horizontal in that specimen. The end purpose for this information is usually some petrographic or geophysical study. For example, the metamorphic petrologist may need to know in which orientation certain crystal axes are aligned. Therefore, the microscope thin section must be cut accurately with respect to geographic coordinates. In paleomagnetism, drill core must be extracted from the rock, in either the field or the laboratory because the specialized equipment needs cylindrical specimens whose orientation is precisely known with respect to north and the horizon.

Figure 5.9(a and b) indicates the way in which a specimen should be labeled so that it may be uniquely reoriented in the laboratory. In the authors' laboratory, such specimens are held in tilting vices to recreate the field inclinations of the marked surface. Since the markings are sometimes partly obscured or erased in transport, it is good practice to make a sketch of the specimen and its orientation mark in the notebook (a digital camera picture may help).

The author prefers to partly loosen the specimen and then mark up its orientation with the compass. Marking a surface first may waste time, since my first choice specimens always seem to disintegrate. Using a chisel and hammer is almost obligatory. Simply bashing an outcrop with a large hammer and hoping to loosen a specimen suitable for careful orientation is usually overoptimistic.

Most paleomagnetists and some petrofabricists prefer to drill core in the field using a diamond-tipped core drill (Figure 5.9(c)). (The drill bits are of stainless steel, which has low remanence and water cooling prevents high temperatures so that paleomagnetic signals are not spoiled.) The advantages are obvious; one can choose exactly the specimen that one wants. The disadvantages are numerous. First, logistically, it is difficult without an assistant since an ample water supply and gasoline must be carefully

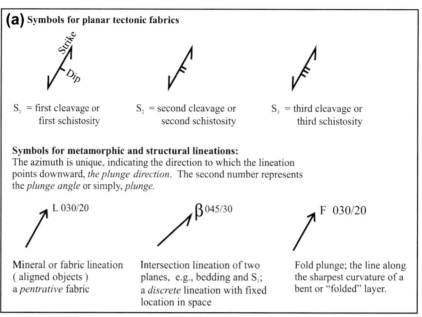

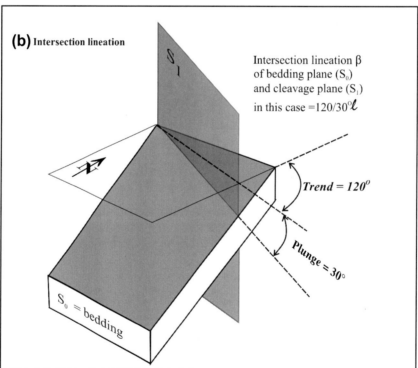

FIGURE 5.8 Largely beyond the scope of this book, lineations may also be recorded.

managed. In many parts of the World, water is quite difficult to find! Drill bits cost currently at least US$100 per piece and may only survive long enough to make a dozen cores in crystalline rock. Scientifically, there are also some difficulties. First, the core sample is quite small and does not allow for many independent paleomagnetic tests and also microscope thin sections. Second, the core is not consistently sized nor is it smoothly cylindrical due to drill vibration. This can be disadvantageous for magnetic work that requires the core to fit precisely into standard equipment holders. Third, it may be difficult to ensure that the core orientation has not been lost due to rotation in the hole. Fourth, there are so many conventions for recording the orientations of core that great care has to be made in choosing a system and using it consistently. Almost every paleomagnetic laboratory has custom software that

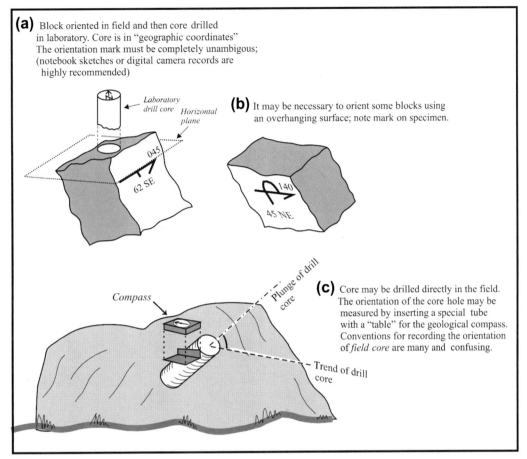

(a) Block oriented in field and then core drilled in laboratory. Core is in "geographic coordinates" The orientation mark must be completely unambigous; (notebook sketches or digital camera records are highly recommended)

Laboratory drill core *Horizontal plane*

045

62 SE

(b) It may be necessary to orient some blocks using an overhanging surface; note mark on specimen.

140

45 NE

Compass

Plunge of drill core

(c) Core may be drilled directly in the field. The orientation of the core hole may be measured by inserting a special tube with a "table" for the geological compass. Conventions for recording the orientation of *field core* are many and confusing.

Trend of drill core

FIGURE 5.9 In specialist work, the compass may be used to retrieve oriented specimens of rock.

requires these orientation conventions to be presented in a unique manner. A fifth and final disadvantage is that every site has core in a different orientation. After some time, it may be difficult to retrieve the documents that report their orientations.

The chief advantages of oriented blocks (Figure 5.9(a and b)), as used in the authors laboratory, are multiple cores, simple orientation coordinates (each core top is horizontal with a north arrow), and perfectly cylindrical laboratory-drilled core. After 30 years, I could retrieve cores from my collection and know their orientation without tracing documentation.

READING PUBLISHED GEOLOGICAL MAPS

Alston Map Area, NE England

It is now time to inspect some more published maps and to apply the simpler elements discussed in the previous pages (Figure 5.10). Consider the Map of the Alston Area of NE England. Primarily, nearly horizontally bedded, Carboniferous sedimentary rocks are exposed on flat-topped moorland hills cut by deep valleys in which alluvium and glacial

boulder clay have accumulated. Consider the extract from the map, reproduced below the main map. Note that, like the map of the Grand Canyon, horizontal and gently dipping strata follow contours and yield a dendritic pattern. In the lower map, at a slightly different scale, a single sandstone stratum is compared with two prominent topographic contours.

Q.1. In the northern part of this map, what is the degree of correlation between the outcrop of the bed and the topography? What does this tell us about the angle of dip of the bed?

Q.2. Below the main map, the details of one horizon, the firestone sandstone formation are extracted. Near locations (a)–(d), mark which side of the bed is the top and which is the base.

Q.3. In the south central part of the extracted map, near locations (a)–(d), does the sandstone horizon run everywhere parallel to the topographic contours? Does the bed tend to drop in a certain direction?

Q.4. The sandstone outcrop shows many "V-shaped" deflections where it crosses stream courses. Using

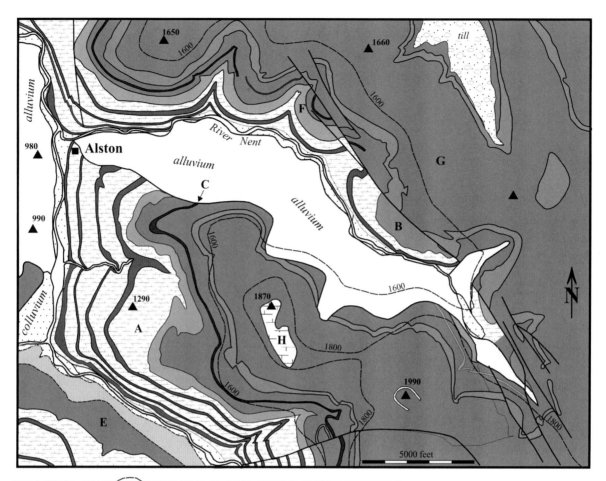

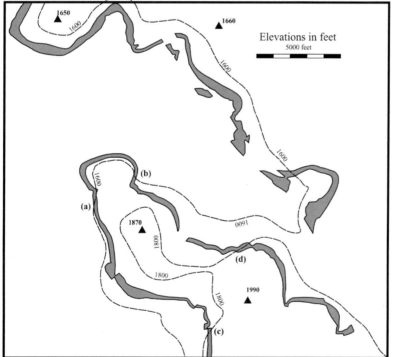

Outcrop of the "Firestone" Sandstone Formation" compared with two significant topographic contours. The heights of some peaks are also shown.

At locations **(a)** to **(d)** mark neatly and clearly which side of the bed is its top and which is its base. Do this in pencil since the map will also be used for a later exercise

FIGURE 5.10   Excerpt from the Alston map, northern England.

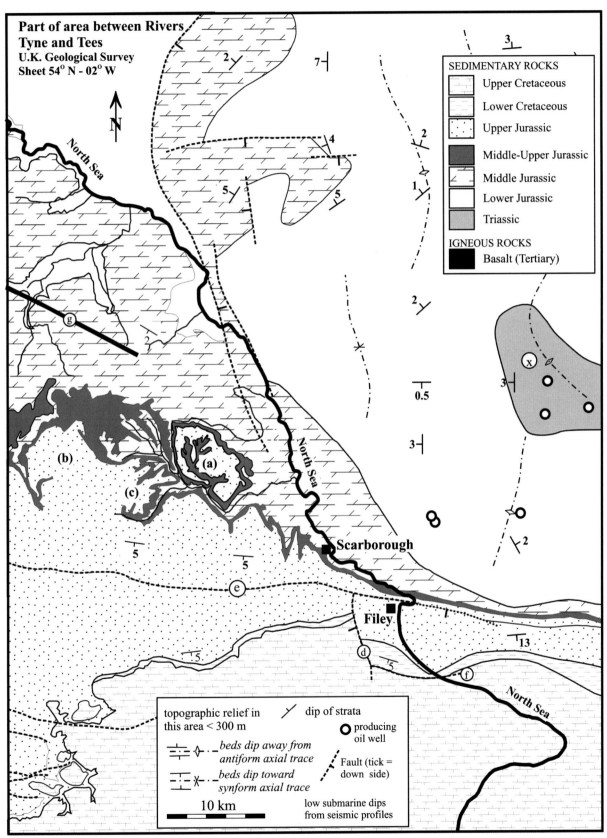

FIGURE 5.11 Onshore and offshore geology near Scarborough, NE England.

that information alone, what could we have inferred about the angle of dip and the direction of dip of the beds? (Remember this inference is always relative to the slope of the stream.)

The area is historically very important for the presence of coal seams, lead–zinc veins, and fluorite deposits.

Q.5. What are the prominent azimuths of the mineral veins (azimuth = trend), shown by the thin black lines? Record your angles in a list. Since they are azimuths, a vein trending 045° could also be reported as 225°. To avoid ambiguity and reduce confusion report azimuths only in the NE and SE quadrants.

Q.6. List some prominent fault trends. Are any faults similar in orientation and location to the veins? If so, which came first, fault or mineral vein?

Tyne–Tees Map Area NE England

This is a much more modern map that includes recent offshore mapping in connection with petroleum exploration (Figure 5.11). Offshore from Scarborough lies an active oilfield; producing wells are shown by a black circle with a dot in the center. This is a large scale map in which topographic relief is not shown by contours. Nevertheless, small topographic variations have controlled the outcrop pattern and you must learn to recognize these patterns. As in the preceding map, dips of strata are rather low or nearly horizontal in some places.

Q.1. At locations (a), the flat-topped moorland is incised by some stream valleys. From the outcrop pattern, what would you infer about the dip of the strata? Indicate this with an appropriate symbol. Is the pattern similar at (b) and (c)?

Q.2. What is the relative age of the faults (d) through (f)? Place them in order, oldest to youngest.

Q.3. What is the shape and orientation of the basalt igneous intrusion (g)?

Q.4. Is there a fault which apparently does not produce any significant movement of the strata? Indicate its location on the map. (Note that the displacement of strata is not simply related to motion on the fault; all the faults in this area have vertical motion, i.e., one side is dropped vertically relative to the other).

Q.5. What possible reasons could explain the fact that the Middle Jurassic outcrop becomes narrower to the SE?

Q.6. Twenty kilometers NE of Scarborough, an area of Triassic sedimentary rocks crops out on the seafloor. Describe how the strata dip in that area. Could this explain why that area hosts a producing oilfield?

Q.7. An outlier is an area of younger rock, surrounded, in map view by older rock. An inlier is an area of older rock surrounded by younger rock. Indicate an example of each on the map. Of the locations marked as (x) and (a), which is the inlier and which is the outlier?

Strata and Plane-Dipping Features

We are very familiar with topographic contours of landforms. Since landforms are complex surfaces, topographic contours (and bathymetric contours below water level) follow contorted paths tracing out a horizontal line on the map. It is often convenient to imagine oneself "walking out contours" and so visualize the land surface and its slopes. The closer contours are spaced, the steeper the surface. Occasionally contours merge, one above the other, where a cliff is located.

Most geological surfaces are much simpler than topographic surfaces and many are essentially planar. Bedding planes for strata of constant dip direction and constant dip therefore have straight "structure contours". Since the dip angle is constant, these structure contours will be equally spaced and parallel. The closer the structure contours, the steeper is the dip of the bedding plane. Note that "structure contour" is preferred to "stratum contour" since the construction may be applied to any geological surface, not just a bedding contact. It is important to remember that the structure contour defines a surface, e.g., the top or the bottom of a bed. Therefore, care must be taken to work consistently only with one geological surface.

In Figure 6.1, we see a similar topographic situation with a sandstone bed cropping out across a hill. Note that the cross-sections have the same vertical and horizontal scale; this is essential to preserve dip angles and bed thicknesses that would otherwise be distorted. In (a), the complete sympathy of the bedding planes with topography indicates that they trace out topographic contours. What is the elevation of the base of the bed (interpolate between mapped topographic contours)? What is the thickness of the

bed, determined from the map? What is the approximate precision of that estimate (i.e., ±how many meters)?

In Figure 6.1(b), the sandstone bed shows no correlation with topography. In map view, it is viewed "edge on" and must therefore be vertical. In this case, the map scale is used to measure its thickness.

In Figure 6.1(c), the bed dips 45° to the east; its strike is N–S. The outcrop pattern shows some correlation with topography. Traced from the south, the bed rises gradually and diagonally across the topographic slope NE toward the hilltop. Then, it traces gradually down hill diagonally to the NW.

The field geologist indicated the dip and strike in Figure 6.1(c) with the appropriate symbol. However, without the field geologist's information, we could determine that dip and strike from the map using a construction that will eventually provide much more information. Consider Figure 6.1(d). First, we must isolate a specific planar surface, e.g., the top or base of the bed. Here, we chose the base. Now, trace the base across the map, clearly marking its intersections with topographic contours.

Since the base of the bed is approximately planar, any two points at the on it at the same elevation must be connected by a hypothetical line of constant elevation. These define the structure contour for the base of the bedding plane. Thus, two points on the base of the bed at 200 m define the structure contour SB200. This may be extended across the map to indicate where the base of the bed is located at a 200 m elevation. Note that the base of the bed in some places has been eroded away, elsewhere, it locates the base

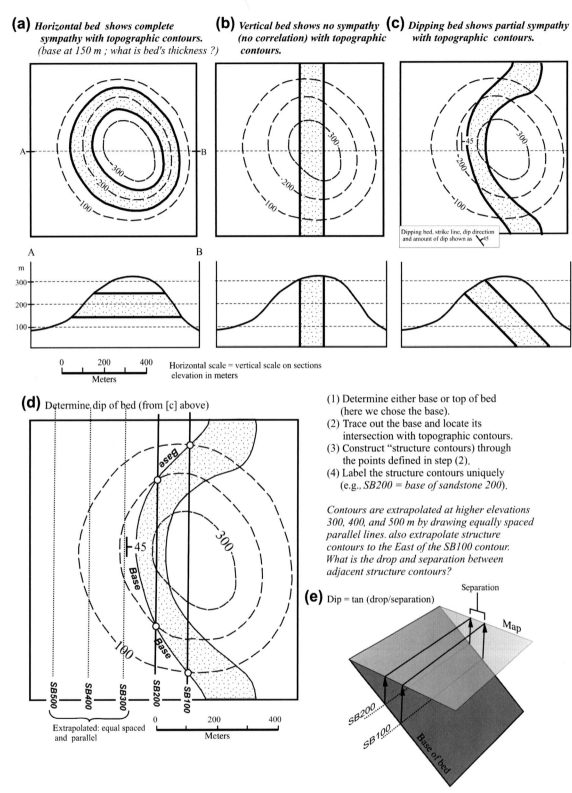

(a) *Horizontal bed shows complete sympathy with topographic contours. (base at 150 m ; what is bed's thickness ?)*

(b) *Vertical bed shows no sympathy (no correlation) with topographic contours.*

(c) *Dipping bed shows partial sympathy with topographic contours.*

Dipping bed, strike line, dip direction and amount of dip shown as ↘45

Horizontal scale = vertical scale on sections elevation in meters

(d) *Determine dip of bed (from [c] above)*

(1) Determine either base or top of bed (here we chose the base).
(2) Trace out the base and locate its intersection with topographic contours.
(3) Construct "structure contours) through the points defined in step (2).
(4) Label the structure contours uniquely (e.g., *SB200 = base of sandstone 200*).

Contours are extrapolated at higher elevations 300, 400, and 500 m by drawing equally spaced parallel lines. also extrapolate structure contours to the East of the SB100 contour. What is the drop and separation between adjacent structure contours?

(e) Dip = tan (drop/separation)

Extrapolated: equal spaced and parallel

FIGURE 6.1 Topographic expression of stratigraphy.

of the bed underground. Similarly, two points on the base at 100 m define the SB100 structure contour. These contours define a very simple geometrical surface, far simpler than any topographic surface. They lie N–S so that the strike of

the bed is N–S (0 or 360°), and the beds drop from the west to the east so we know the dip direction is eastward.

On many maps, which have uniformly dipping beds, you may extend structure contours from a small area to other areas.

Moreover, since the contours are equally spaced and parallel for a plane, we may extrapolate them to different elevations. For example, structure contours are extrapolated for the base of the sandstone at 300, 400, and 500 m in Figure 6.1(d). No outcrop occurs there because the land surface is below the hypothetical plane. The plane could also be extrapolated to the east, with contours at sea level (=0 m), −100 m, −200 m (below sea level). For example, these would predict the locations of the bedding plane at mines at those depths. Draw in those structure contours for practice. You may also interpolate structure contours (for example, at 350 and 450 m) if this helps in the precision of mapping in the geological boundaries.

Return to Figure 5.10 and examine the extracted map below the main map.

Q.1. In the northern part of the lower map, what does the relationship of the firestone sandstone formation with topographic contours tell you about its angle of dip?

Q.2. Consider the southern outcrops of the sandstone, near (a)–(d). Determine the base of the sandstone and construct its structure contours at 1600 and 1800 feet. Are the structure contours parallel? Discuss the result with your instructor. Because the beds here dip gently and they are not precisely planar, structure contours are neither straight nor parallel. This situation is very realistic but it does not devalue the use of structure contours.

STRUCTURE CONTOURS: CONSTRUCTED, EXTRAPOLATED, INTERPOLATED, AND TOPOGRAPHIC INTERSECTIONS

We learned how to construct structure contours from the intersections of a geological plane and topographic contours. Two points on the plane at the same elevation suffice to define a structure contour line. Due to variations in strata dips, mapping errors, etc., it is better to use more than two points at the same elevation, if they are available.

For a uniformly dipping plane, structure contours are parallel lines, equally spaced, for higher and lower elevations of the geological plane surface. This is extrapolation of structure contours. However, sometimes the spacing of available structure contours is insufficient to solve a particular problem. Therefore, one may interpolate structure contours between those constructed from the base map. For example, the base of sandstone may have structure contours at 100 and 200 m but you may draw in an intermediate parallel line to define the 150 m structure contour. This requires the "estimation" of the 150 m topographic contour but it is a useful method to complete some more complex map boundaries.

Finally, there is a paramount consideration concerning the intersection of a geological contact with a topographic contour. A geological boundary may only occur where it is justified by the intersection of a topographic contour and stratum contour of the same elevation. Doubt may creep in where the frequency of topographic and structural contours is too low, i.e., they are spaced too far apart. Interpolating and sketching in extra structure and topographic contours usually resolves this snag. For example, if the published contour interval is 100 m and that leaves you with some doubt in constructing a map, interpolate some contours at 50 m intervals.

DETERMINING DIP ANGLES FROM STRUCTURE CONTOURS

If a plane-dipping surface, such as the base of a bed, was exposed, we could draw horizontal lines on it, at elevations of say 100, 200, and 300 m. These are structure contours. Since the surface is plane these would be straight, their direction defines the bed's strike. Since the slope is uniform, they would be equally spaced in plan view.

If the slope was vertical, the structure contours would appear to lie directly upon one another in map view, as with topographic contours on a cliff. Their separation would be zero. If the slope was very gentle, the separation of the structure contours in map view would be very large. The greater the separation of the structure contours, the smaller the dip. If the 100 m contour and the 200 m contour are separated by 100 m in map view, the dip is defined as

$$\text{Tangent (dip)} = \text{drop/separation}$$
$$= 100/100 \text{ thus, the dip angle is } 45°.$$

The drop is the difference between any pair of structure contours and the separation is the distance between them in map view. (Using nonadjacent structure contours, with a greater separation and greater drop may give a better average value by permitting small drafting errors to cancel out.)

In most cases, we do not need to use a calculator. Table 6.1 converts the ratios of drop/separation into dip angles. The accuracy of geological field measurements and their intrinsic variability rarely justify better precision than this.

Very small dip angles cannot be measured with a geological compass and reported on field maps. However, in some circumstances, geologists need to know those slopes and slope directions. For example, in sedimentology and tectonic studies, it is important to know the precise values of the very gentle (<1°) slopes of the continental shelves and sedimentary units deposited on them, the very gentle gradients of the basement in continental interiors which control gravity gliding and depositional basins. Also for global tectonics, it is important to know the precise gradients of major crustal or lithospheric discontinuities. Such small dip angles are determined by calculation from geophysical data, usually from seismic surveys or seismological studies and they are expressed as ratios.

Table 6.2 illustrates the small angles associated with such low gradients.

TABLE 6.1 Steep Slopes

Ratio (tangent)	Angle (degrees)	Ratio (tangent)	Angle (degrees)
0.00	0	1.19	50
0.09	5	1.43	55
0.18	10	1.73	60
0.25 (1/4 slope)	14	**2.00** (2/1 slope)	63
0.27	15	2.14	65
0.36	20	2.74	70
0.46	25	**3.00** (3/1 slope)	72
0.50 (1/2 slope)	27	3.73	75
0.58	30	**4.00** (4/1 slope)	76
0.70	35	5.67	80
0.84	40	11.4	85
1.00 (1/1 slope)	45	∞	90

Note to engineering students: Civil engineers determine the gradients (slopes) of roads using the drop and the distance measured along the ground slope, not in plan view. They replace the horizontal separation with the hypotenuse. Thus, they define their slopes using the sine function. For small angles (slopes <25°), the differences between the answers given by these two calculations are minor, in the context of the precision of geological measurements.

TABLE 6.2 Low Slopes

Slope (1 in ...)	Slope	Dip angle (degrees)
10,000	0.0001	0.006
5000	0.0002	0.011
1000	0.001	0.057
500	0.002	0.115
100	0.010	0.573
57	0.017	1.000
50	0.020	1.146
29	0.035	2.000
19	0.052	3.000
14	0.070	4.000
11	0.087	5.000
10	0.100	5.710

EXTRAPOLATING GEOLOGY AND DETERMINING DIPS USING STRUCTURE CONTOURS

At this stage, it is valuable to consolidate the principles accumulated to perform a routine procedure used by field geologists and explorationists. Given geological information from a small area, we shall extend the mapped boundaries, under the assumption that the geological boundaries keep their same orientation across the map.

Examine Figure 6.2(a); the bed shows complete sympathy with topography and therefore must be horizontal.

Q.1. The map shows part of a shale bed; the map is incomplete. From its degree of correlation with topographic contours, extend the horizon across the map, carefully maintaining correlation with the paths of topographic contours. This requires more care and thought than most people realize initially. Your instructor should help with this.

Q.2. Which is the base and which is the top of the shale? Indicate this on the map.

Q.3. What is the thickness of the shale and how precise is that estimate?

Examine Figure 6.2(b), in which case the bed is not horizontal. There are sufficient outcrops to define 600 and 700 m structure contours as shown on the map.

Q.1. Did the geologist draw the structure contours for the base or for the top of the shale?

Q.2. Determine the dip angle from the drop and separation. Mark this neatly on the map using the appropriate dip-and-strike symbol.

Q.3. Extrapolate structure contours across the entire map, carefully labeling them.

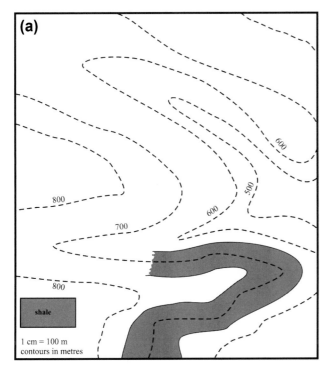

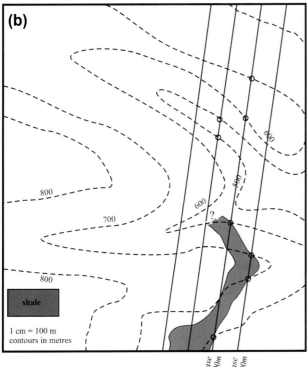

FIGURE 6.2   (a) Horizontal bed. (b) Dipping bed. See text.

Q.4. Identify where each structure contour intersects a topographic contour of the same value. Indicate these points with small circles or dots.

Q.5. Extrapolate the outcrop of the base of the sandstone across the map, joining the points identified in Q.4 but maintaining careful correlation with topography.

Remember, a geological boundary may only occur where it is justified by the intersection of a topographic contour and stratum contour of the same elevation. The simple statement is more difficult to practice.

Q.6. Complete the outcrop of the top of the shale in the same manner.

Figure 6.3(a) shows a very large-scale geological feature with a prominent dip which may be determined from a type of structure contour. You are aware that lithospheric plates are created at mid-ocean ridges (Figures 3.2 and 3.3) and spread sideways under the influence of gravity at speeds of several centimeters per year. Along parts of some margins of some oceans, the cooled lithospheric plate, several hundred kilometers thick, descends ("subducts") under its own weight. Such subduction zones are the sources of deep earthquakes, down to depths at which temperatures suppress earthquake-type failure. The depths of the earthquakes are determined by seismological studies and the map shown also give contours of the average depth to the earthquake focus. The contours are thus structure contours for the "earthquake-producing surface" or Benioff zone.

Q.1. From the drop and separation of contours along the line AB, what is the angle of dip of the Benioff zone?

Q.2. Along CD, a complication occurs. Starting from D, the first few structure contours are in order, i.e., successively deeper contours are met in correct sequence. Can you explain the complication that occurs closer to C?

Figure 6.3(b) provides an exercise in completing the mapped boundaries across the map of a small area, assuming that its dip and direction of dip remain constant. Draw structure contours for the base and for the top of the bed, being careful to distinguish them. Determine the dip and show this appropriately on the map, using the dip-and-strike symbol. Also, draw a true-scale cross-section from A to B with A on the left. A true-scale section has equal vertical and horizontal scales.

COMPLICATIONS IN DRAWING CROSS-SECTIONS: VERTICAL EXAGGERATION AND APPARENT DIP

Vertical Exaggeration

In the previous question, you were specifically asked to draw a true-scale cross-section, i.e., one that has equal vertical and horizontal scales. There are two main reasons for this, arising from the distortion that arises when geologists exaggerate the vertical scale. Usually, large-scale maps exaggerate the vertical scale so that the reader can identify topographic features such as valleys and summits. However, the distortion of the geology may outweigh those advantages. Consider Figure 6.4(a). It is evident that exaggerated vertical scales make structures appear steeper than they are. Moreover, they distort the thickness of

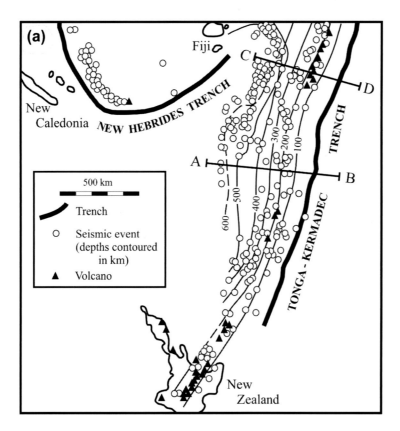

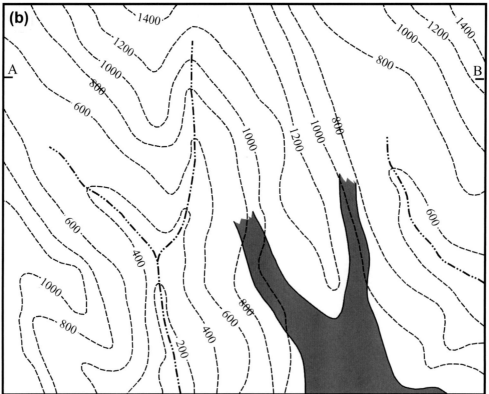

Scale: 1 cm = 100 m or 1:10,000

FIGURE 6.3   (a) Dipping Benioff (subduction) zone. (b) Dipping bed, complete the map.

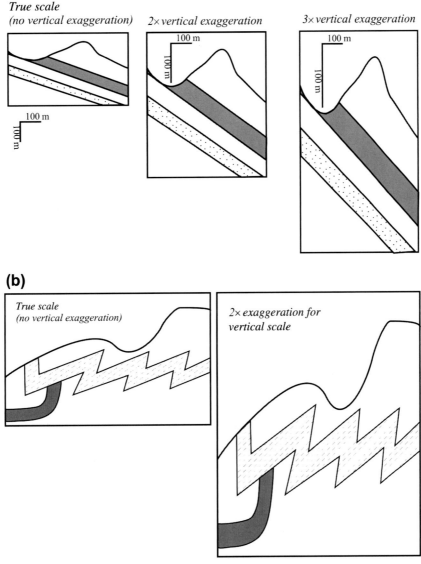

(a)

*True scale
(no vertical exaggeration)*

2× vertical exaggeration

3× vertical exaggeration

(b)

*True scale
(no vertical exaggeration)*

*2× exaggeration for
vertical scale*

FIGURE 6.4   The effects of vertical exaggeration.

layers; steeper layers retaining their thickness, whereas gentler dipping layers appear to be thickened.

Furthermore, these distortions are not linearly progressive if folded layers are present (Figure 6.4(b)). Their dips will be strangely distorted and bed thicknesses vary in meaningless counterintuitive ways. Overall, it is far better to retain true-scale sections wherever possible.

Apparent Dips and True Dips

A further complication from sections through geological maps is that a given section cannot reveal the true dip of every feature. There would be no problem where all the faults and all the beds dipped in the same direction. One could draw the cross-section parallel to the dip direction

(perpendicular to strike) and the true angles of dip would appear in the section. Such a section is a profile section.

In general, the section or plane on which the geology is viewed will cut through the layers revealing an apparent dip that is always less than the true dip. This is a wider problem in geology that confronts any attempt to section rocks, e.g., for microscope work and is sometime referred to as the cut affect.

For our purposes, it is sufficient to note that one cannot take true dip angles from maps and simply draw beds on sections at the same angle, unless the section is parallel to the bed's dip direction. We shall see later that there is a nonmathematical technique for neatly sidestepping this problem but Figure 6.5(a) shows the trigonometrical relationship.

(a) d = true dip of bedding

α = apparent dip of bedding on some arbitrary vertical plane

(b) Nomogram of relationship

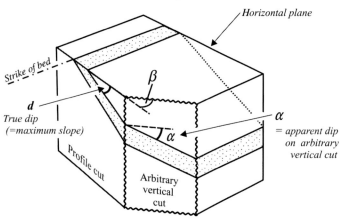

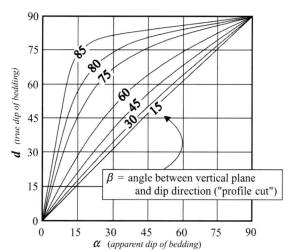

The construction refers to the true and apparent dip of a bedding plane; of course, it is equally applicable to any geological plane such as a fault or fold-axial plane.

(c) "Rake" of the arbitrary vertical cut on bedding

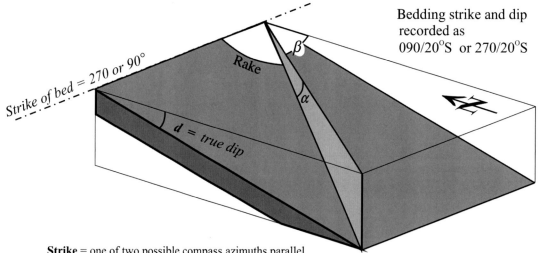

Bedding strike and dip recorded as 090/20°S or 270/20°S

Strike = one of two possible compass azimuths parallel to a horizontal line on the bed

d = true dip of bed, in vertical plane perpendicular to strike of bed

α = apparent dip on arbitrary vertical plane that makes angle (90-β) with strike of bed; (β is measured in the horizontal plane).

Rake > (90-β) is the angle **on the bedding plane**, between the arbitrary vertical plane and bed strike. This uncommon term is describes *the angle between any line and the strike of a plane in which it lies*. One must specify from which end of the strike we measure rake.

FIGURE 6.5 Real and apparent dip.

In a vertical section that makes an angle β with the dip direction (90-β with the strike direction), the apparent dip angle (α) in the section is related to the true dip of the plane (d) by

$$\tan \alpha = \tan d \cdot \cos \beta.$$

Figure 6.5(b) is a nomogram, or graph, that simply relates these angles. In Figure 6.5(c), the relationships between true and apparent dip are shown in a manner that may be easier to grasp for some readers. However, the term "rake" is introduced there for completeness and future reference. It refers to the angle between the strike of a dipping plane and a line in the plane, the angle being measured on the dipping plane.

DIPPING STRATA, UNCONFORMITIES ON PUBLISHED GEOLOGICAL MAPS

Using the more elementary information in this chapter, you may now study the following maps. At this stage, we shall not be dealing with structure contours and geometrical constructions. Instead, we shall consolidate what we have learned in some real situations.

Figure 6.6 shows the simplified geology of part of the Isle of Wight, SE England.

Q.1. The legend is divided into two parts, "solid or bedrock geology" and "drift or superficial deposits", tabulated as Recent and Pleistocene. More accurately, superficial deposits are normally termed colluvium. Given the age

of the superficial deposits what is the youngest possible age and the oldest possible age for the land surface on which they are all deposited?

Q.2. In the superficial deposits, one lithology is marked "?". Examine the map; what term would most generally describe this sediment?

Q.3. Trace the river channel south from Yarmouth toward the location marked as "C". Can you explain why the deposit marked "?" extends to the edge of the steep cliffs near C? (Hint: the Cretaceous limestone is very soft.)

Q.4. Draw a sketch cross-section showing the principal formations and their dips, along the section line LR, with L on the left. Note that you may omit topographic relief and assume the terrain is essentially horizontal. Use dip-and-strike information to the east and west of the section line and project it into your section. This is valid since

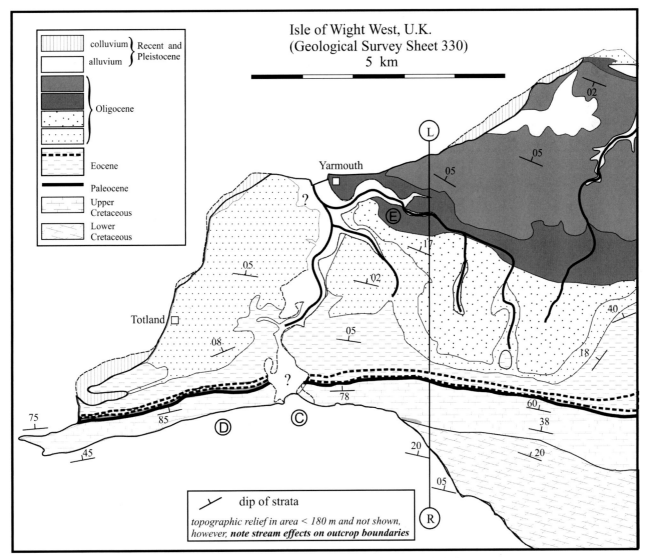

FIGURE 6.6 Simplified geology of western Isle of Wight, England. See text.

the section is perpendicular to strike. The dip angles on the map will therefore be valid if shown directly on the section. (There is no apparent/true dip problem.)

In drawing this cross-section, you only have dips at the surface. You must extrapolate the bed below the surface but they cannot cross one another, of course. This will require you to gradually change the dips of underground beds to conform to those above. In so doing, you will construct the profile of a knee-shaped bend in the strata called a monoform. Discuss this with your instructor.

This simple fold is an expression of the Alpine mountain building event, which just reached the southern shores of Britain at this one location.

Q.4. In which orientation would you expect the beds to young at locations D and E? Indicate this on the map with the preferred younging symbol (Y).

Q.5. Where is the bedrock most nearly horizontal, as deduced from outcrop patterns?

Indicate this location with the appropriate symbol for horizontal strata.

Figure 6.7 shows a very simplified outline of the geology of the Lake District, NW England.

The mountains in this area reach 3000 feet in the SW and around 2000 feet in the NE, separated by the broad, north–south valley of the River Eden. The valley is fault bounded and contains younger sediment in the center. This structure is a rift valley. The topographic relief is rather subdued compared to the scale of the map so that, once again we may regard the surface as essentially flat. We can apply the observations about geological contacts and dip-and-strike symbols to understand the geological events in this region.

Q.1. How many unconformities can you identify? State what criteria were used in their identification and mark them on the geological column.

Q.2. Which unconformity is the most significant one?

Q.3. Is there evidence for folding beneath any unconformity?

Q.4. What is the age of the granite intrusions relative to the adjacent formations?

Q.5. Examine the intrusions formed by the diabase ("dolerite" in local terminology)? Which term describes the diabase starting just south of Carlisle and tracking across the Eden Valley in an NW to SE direction at (b)? What term best describes the dolerite intrusions that occur in N–S directions along the east side of the Eden Valley (at (a))? What are the relationships of these diabase intrusions to bedding?

Q.6. What are the upper and lower limiting ages for the diabase intrusion?

Q.7. Just west of Carlisle, an outlier of Jurassic strata crops out. The eastern part of the outcrop is bounded by a fault (c); which side of the fault is downthrown? On smaller scale maps, the downthrown side of a fault is shown with a tick, as on the Tyne–Tees map (Figure 5.11). Indicate the downthrown side of the fault with a small tick.

Q.8. Examine the major NNW–SSE trending faults which bound the Eden Valley. Which sides are the downthrown sides? Indicate them with neat small ticks. Discuss with your instructor if the term rift valley is appropriate for the Eden Valley.

Q.9. The sedimentary rocks in the Eden Rift Valley dip to the ENE. Does this have any implications for the relative magnitude of the downthrow of the faults on either side of the valley?

Q.10. Folds are bends or warps in strata; antiforms are arches or upwarps (see Figure 5.11, Tyne–Tees area) and synforms are trough-like bends in strata. Can you identify a synform near Carlisle and an antiform in the Ordovician rocks? Mark their trend using the appropriate symbol shown in the legend.

SIMPLE CONSTRUCTIONS FOR MAPPING OUT STRATA FROM A FEW OUTCROPS

How do we create published maps? How does the geologist, from a few scattered outcrops determine the course followed by bedding contacts across hillsides and valleys? It is very rare that contacts are sufficiently well exposed for a geologist to walk along them and draw their track on a map. We can make reliable geological maps where less than a fraction of 1% of the land surface is exposed bedrock.

A few simple geometrical concepts permit the beginning geologist to extend information for planar beds from individual outcrops to complete a map. At a later stage, the geologist learns to adapt the techniques for the planar parts of structures such as folds that cause strata to be nonplanar on a larger scale.

Three-Point Construction

Consider the very simple situation in Figure 6.8(a). Three outcrops expose a bedding plane. It is very rare that two or more outcrops will crop out at the same elevation. Thus, some work is required to define the trend of the first structure contour. The outcrops expose the bedding plane at 150, 100, and 300 m. Since a plane has a uniform gradient in any given direction, we know that at some point between B (100 m) and C (300 m), the plane must be at 150 m. If we can find that point, we may join it to point A where the plane crops out at 150 m. The 150 m point between B and C must lie closer to B than C and length ratios determine its position (Figure 6.8(b)). The point must be located (BX/BC) of the distance from

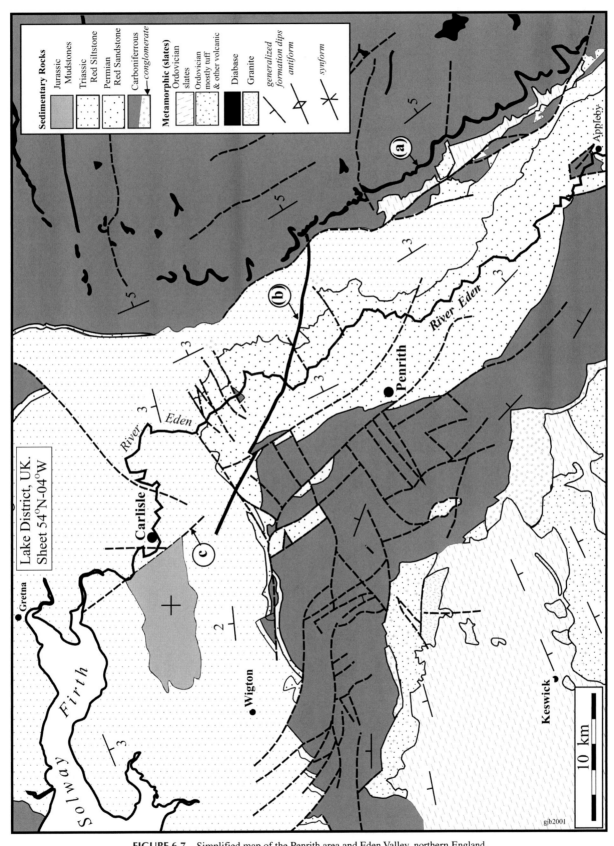

FIGURE 6.7 Simplified map of the Penrith area and Eden Valley, northern England.

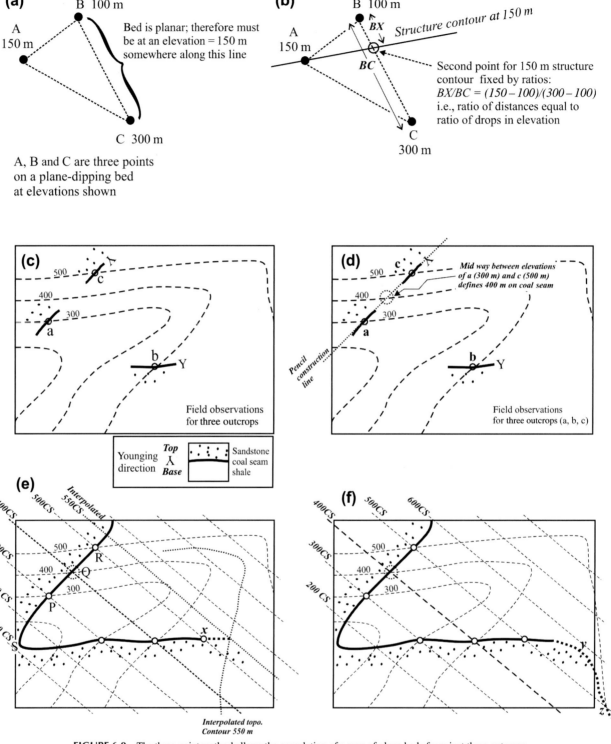

(a) B 100 m

A
150 m

C 300 m

Bed is planar; therefore must
be at an elevation = 150 m
somewhere along this line

A, B and C are three points
on a plane-dipping bed
at elevations shown

(b) B 100 m

A
150 m

Structure contour at 150 m

BX

BC

C
300 m

Second point for 150 m structure
contour fixed by ratios:
BX/BC = (150 – 100)/(300 – 100)
i.e., ratio of distances equal to
ratio of drops in elevation

(c)

500

400

300

a

c

A

b Y

Field observations
for three outcrops

(d)

500

400

300

a

c

A

*Mid way between elevations
of a (300 m) and c (500 m)
defines 400 m on coal seam*

b Y

*Pencil
construction
line*

Field observations
for three outcrops (a, b, c)

Younging
direction

Top
Λ
Base

Sandstone
coal seam
shale

(e)

400CS 500CS 550CS *Interpolated*

300CS

200 CS

100 CS S

500

400

300

R

Q

P

x

*Interpolated topo.
Contour 550 m*

(f)

400CS 500CS 600CS

300CS

200 CS

500

400

300

Y

FIGURE 6.8   The three-point method allows the completion of a map of plane beds from just three outcrops.

B toward C. That ratio is (150 – 100)/(300 – 100) or ¼
of the way from B toward C. Once that point is defined,
the first structure contour is drawn and the others may
be drawn parallel to it through A and through C. This
is the three-point construction; the three points may be

surface or borehole information with at least two different elevations.

In some instances, it may be necessary to use the drafting technique shown in Figure 6.9 to extrapolate suitably
spaced structure contours and determine their spacing.

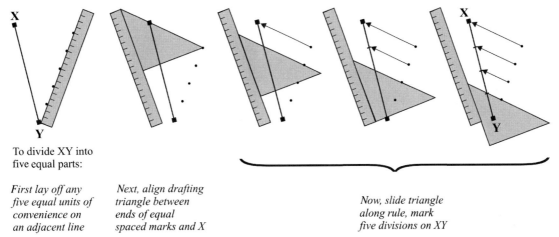

To divide XY into
five equal parts:

*First lay off any
five equal units of
convenience on
an adjacent line*

*Next, align drafting
triangle between
ends of equal
spaced marks and X*

*Now, slide triangle
along rule, mark
five divisions on XY*

FIGURE 6.9 Technique for drawing parallel strike lines. Note in advanced problems, strike lines may at best only be approximately parallel.

Extending Geological Boundaries by Construction

In some earlier exercises, you learned intuitively how strata must crop out systematically in relation to topography; we used the terms topographic sympathy and topographic correlation to express this concept (Figures 5.1–5.3, 6.1, and 6.2). Figure 6.10 shows the partial outcrop of a shale–limestone contact in two large outcrops with the bedding plane contact cropping out at three different topographic contours. Structure contours are drawn for you, so that you may reproduce them on the solution page (lower part of Figure 6.10). The stratigraphic column shows the sequence determined from boreholes. The limestone and shale have vertical thicknesses of 50 m, whereas the thickness of the shale and marl are unknown because either the top or the base is absent in the borehole. You may assume that the beds dip uniformly and you will discover that the dip is sufficiently gentle for you to assume that the vertical thicknesses are good approximations to the true bed thicknesses. Since you know the vertical bed thickness is 50 m, the structure contour for one contact may be relabeled with a different elevation for another bedding plane.

From the outcrop of the bedding plane between the shale and limestone, complete the geological map of the area using Figure 6.10. Note that topographic contours are in 50 m intervals.

Proceed as follows: First determine whether the top or the base of the limestone is shown, then use a three-point construction to construct structure contours for the limestone–shale contact. Extrapolate the structure contours across the map. From their intersections with topographic contours of the same value, map out the shale–limestone contact. Use the thickness of the shale and limestone to complete the map of the shale and limestone beds.

What is the depth to the limestone–shale contact at *x*? What are the maximum possible thicknesses for the shale and marl in this area?

Figure 6.11(a) shows the base of a carbonate iron formation traced across part of the map at its eastern edge. Note that contour intervals are 50 m. The area shows a topographic depression, the ticks on its topographic contour indicate that this is a depression rather than a hill. Closed depressions are rare on topographic maps. In this case, the depression is due to subsidence over a solution cavity in soluble rock, called a sinkhole. (Such sinkholes are common over limestone. Other causes of closed topographic depressions are various relatively rare glacial processes, volcanic, and impact craters.)

Construct structure contours for the base of the carbonate iron formation; not all of the base is shown so you must map in another small piece of it. Assuming that the vertical thickness of the iron formation is 50 m, you may relabel the structure contours for the base and use them to map in the top of the iron formation. The upper map shows the structure contours for the base of the formation. Determine the borehole log in a vertical borehole at *X*.

Figure 6.11(b) shows the same are after open-pit mining in the south central portion of the map. Remap the geological boundaries, taking into account the slightly revised topography.

Figure 6.12 is a simple revision of these ideas. The bedding plane crosses three topographic contours, which provides sufficient information to complete the map. If in doubt, draw one or more cross-sections. You may also interpolate approximate structure and topographic contours, for example, at 50 m intervals, between the 100 m intervals given.

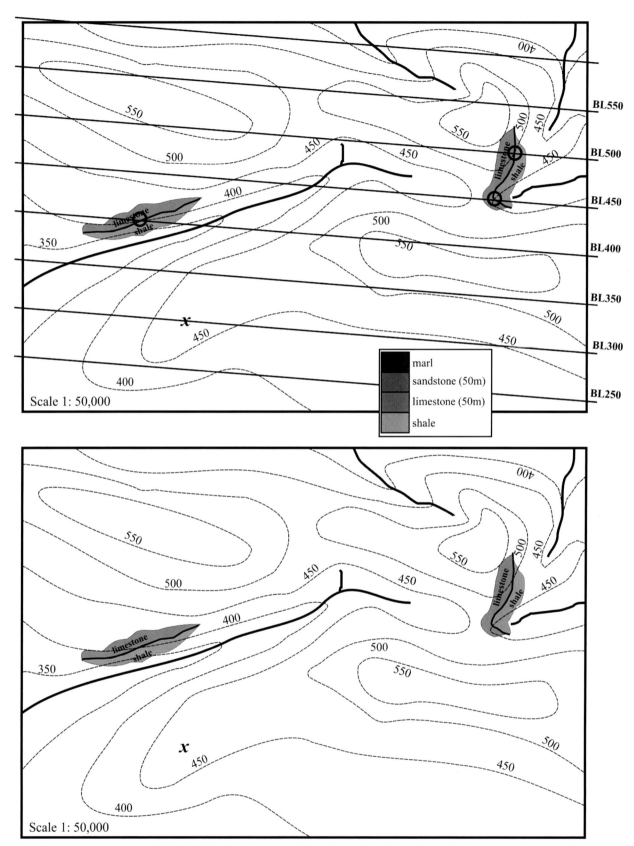

FIGURE 6.10 Complete map of plane-dipping beds (three-point problem).

(a)

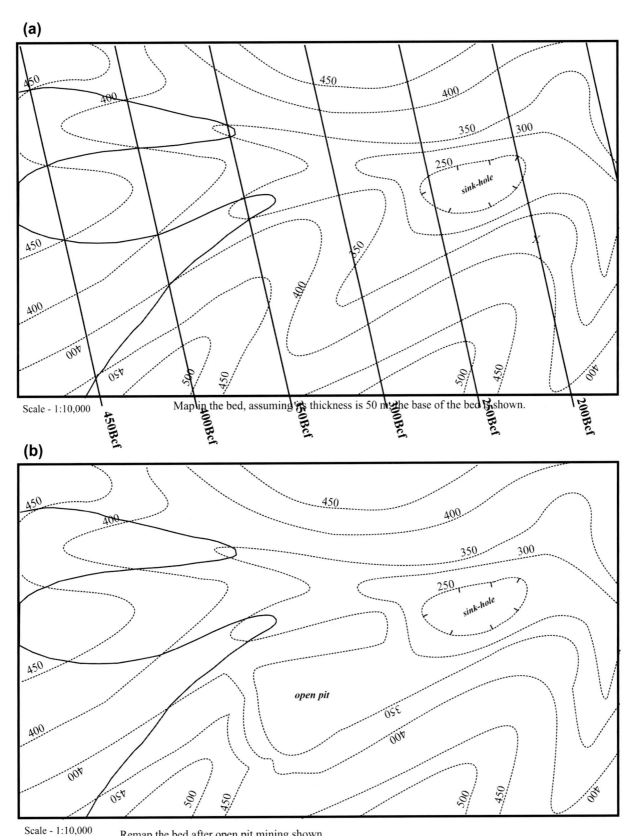

Scale - 1:10,000   Map in the bed, assuming its thickness is 50 m; the base of the bed is shown.

450Bcf   **400Bcf**   **350Bcf**   **300Bcf**   **250Bcf**   **200Bcf**

(b)

Scale - 1:10,000   Remap the bed after open pit mining shown.

FIGURE 6.11   (a) Base of bed is shown; map in its top assuming that it is 50 m thick. (b) Remap the area after the open-pit mining shown.

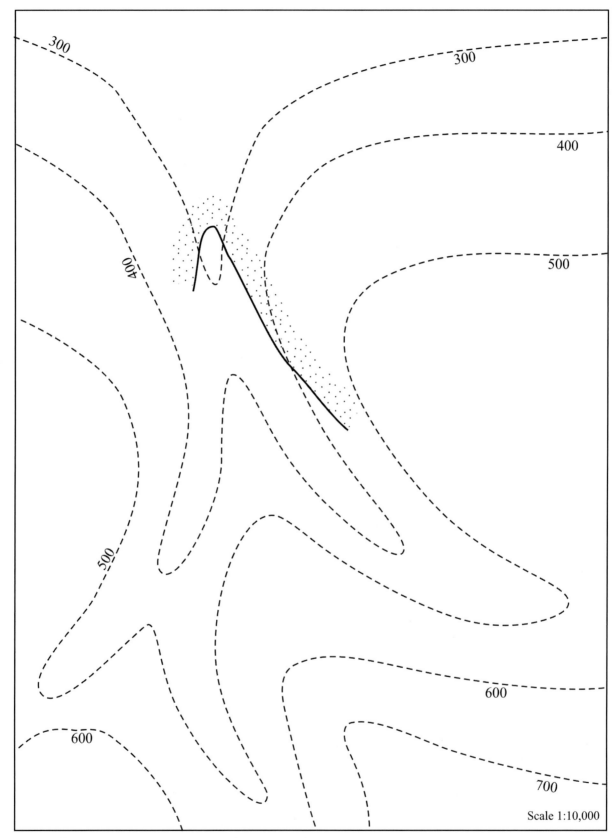

FIGURE 6.12 Complete the map (three-point problem). Indicate the dip and strike of the beds and show which bed is older.

Dips, Thicknesses Structure Contours and Maps

Previously, we learned that a geological surface may be contoured as though it was a topographic surface. However, most geological surfaces are rather simple and many are planar. Even where they are not planar surfaces they may often be decomposed into planar parts that we investigate separately, for the sake of geometric convenience. Let us consider the planar element, for example, a bedding plane, further.

A horizontal line drawn on the plane is a strike line (its direction defines the "strike" of the feature). This simple straight line, for example, at 200 m elevation on a bedding plane would be described as the 200 m structure contour for that bedding plane. We must be very careful to work with a unique bedding plane, either the base or the top of the bed. At some time, we may need to construct structure contours for both the top and the base but labels must appropriately distinguish their structure contours. Follow the instructions closely; although elementary, this is a key exercise.

Figure 7.1 shows two series of strata; E and F overlie C and D unconformably. The unconformity is the base of E. You may verify this by imagining yourself "walking" NW up the valley; obviously one meets the base of E first. A little higher up, we find the top of E, in contact with F. Thus, the plane EF defines the top of bed E. The aim of the exercise is to draw a cross-section AB. However, one cannot simply transfer the dip of beds into the section because they do not in all cases dip parallel to the line AB. In other words, some construction must be used to put the correct apparent dips (Figure 6.5) in the plane of section AB.

Figure 7.2 shows the procedure incrementally. Follow the instructions and then return to Figure 7.1 to complete the section for yourself. Consider the top of bed E in Figure 7.2. It occurs in two places at 600 m. Therefore, you may construct a stratum contour EF600 across the map. Where this meets the line of section AB permits us to locate it at the correct elevation (600 m) in the section at the bottom of the figure. Next, we identify another elevation at which bedding plane EF meets a contour on the map. This is at 700 m permitting us to construct a 700 m structure contour EF700. We can extrapolate this into the cross-section. Now, in the cross-section, we have two points for the bedding plane EF, at 600 and 700 m. Connect these in the section to draw an accurate cross-section of bedding plane EF. Note that it was not necessary nor would it have always been correct to use the dip from the map to construct a section. The bedding plane below the unconformity, CD, illustrates the importance of using structure contours to determine the correct apparent dip in the any general section.

Note one very important philosophical concept here. We draw structure contours right across the map. For example, it implies that on the northern edge of the map and on the western edge of the map, we will find EF at 600 m, underground. This indicates the important predictive value of structure contours. We know how deep we must drill or mine to find EF at different parts of the map where it is not exposed.

Understanding Geology Through Maps. http://dx.doi.org/10.1016/B978-0-12-800866-9.00007-7

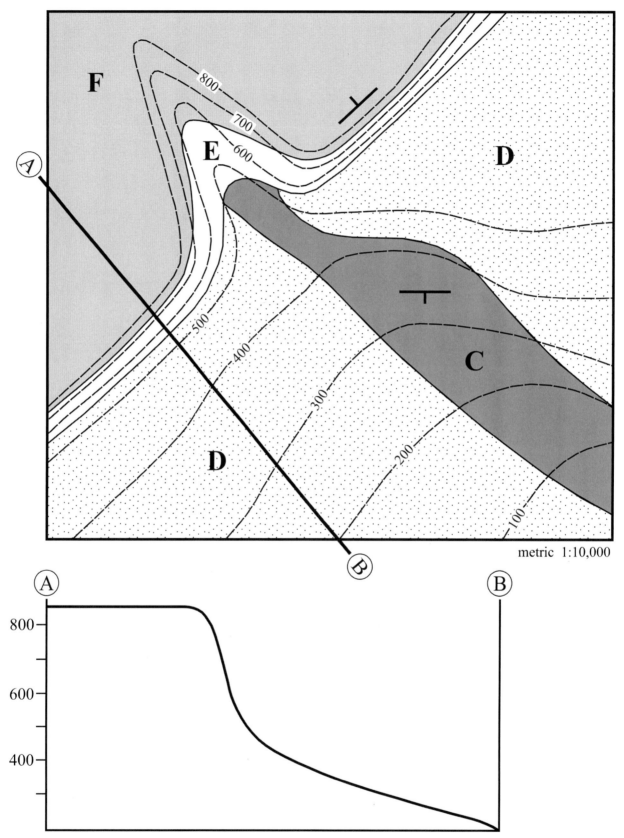

FIGURE 7.1 Draw an accurate cross-section AB that takes account of apparent dip affects by using the construction shown in Figure 7.2 and text.

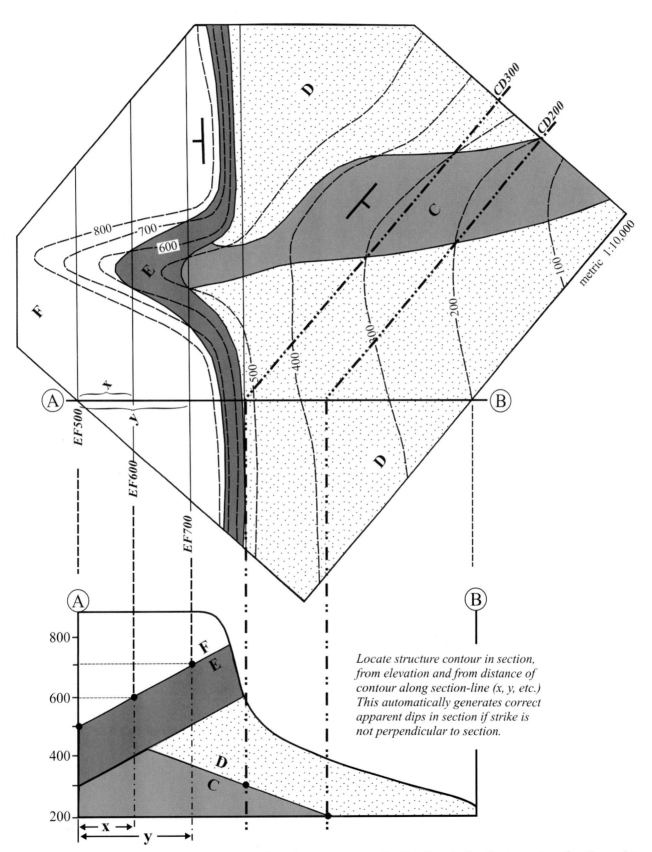

FIGURE 7.2 The previous map and the necessary construction to draw a true cross-section. Note the projection of stratum contours from the map into the section.

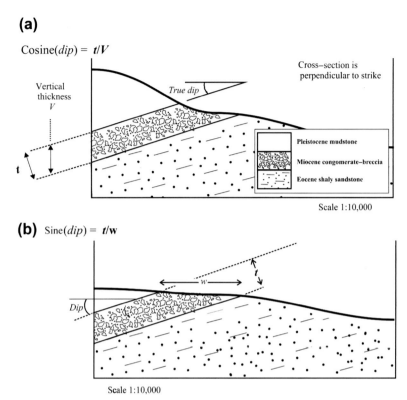

FIGURE 7.3 (a) Determining true (*t*) and vertical (*V*) thickness. The two differ slightly if the beds dip gently (e.g., <20°). (b) For areas with low topographic relief, the outcrop width may be used to estimate layer thickness.

Usually of less practical value, but important geologically, we note that in the center of the valley, very close to the letter "E", the land surface is lower than 600 m. Nevertheless, we have drawn the structure contour there. This tells us where the EF600 level used to be, before erosion by the stream course. It indicates how much material has been eroded.

We may extend the procedures across the entire map to complete the section. We calculate dip angles for the map (true dips) from the separation of structure contours using

$$\text{Tangent (dip)} = \text{drop/separation}$$

for which Table 6.2 gives suitably accurate answers.

The drop between EF600 and EF700 is 100 m, by definition. Determine their separation, in meters, knowing that the scale of the map is 1 cm = 100 m. With drop and separation, both, in meters, calculate the dip. Write the dip angle against the tick part of the dip-and-strike symbol.

Repeat the above steps for the contact CD. Note that the base of C is not exposed but the CD contact must be the top of bed C because the rock C occurs below the CD contact, in the center of the valley.

There are no short cuts to drawing the cross-section. Follow the recipe exactly and it will prevent future calamities,

especially where lines of sections are not parallel to the dip direction. In fact, even in this map, the sedimentary rocks dip in different directions so that no section can reveal the true dips of both sequences. Thus, the angles of beds, etc., in cross-sections are rarely the true dips and apparent dips are always less than true dips. This is also the case when we view strata on fortuitous outcrop surfaces in the field. Complete the section in Figure 7.1 using these techniques.

TRUE THICKNESSES OF DIPPING BEDS

The true thickness of a bed is defined as the distance between the top and the bottom, in a direction perpendicular to the stratification. This is the true thickness or orthogonal thickness (*t*) (Figure 7.3). However, in vertical sections, dipping beds exaggerate the true thickness (*t*) to some larger value, *V*, the apparent vertical thickness as measured in vertical boreholes. The latter is usually not of any fundamental importance to the geologist so that we must calculate *t* from *V*, knowing the true dip.

Method (1): True Thickness from Vertical Thickness in Borehole

Where beds dip horizontally, thickness is measured as a vertical separation, for example, in a cliff or in a vertical borehole. (If the beds dip vertically, bed thicknesses are measured as a horizontal separation from the plan view given on a geological map.)

However, in general, beds dip at some angle. Thus, measurements from steep boreholes will be misleading, overestimating the orthogonal thickness of the bed in proportion to the dip angle. Although, this vertical thickness (V) exaggerates the thickness of nonhorizontal beds, we may use it to estimate the true or orthogonal thickness (t) if we know the dip angle (see Figure 7.3(a)):

$$\cos\,(dip) = \frac{t}{V} \text{ so that } t = V \cdot \cos\,(dip).$$

Method (2): True Thickness from Vertical Separation of Stratum Contours

Stratum contours may also be used to determine the vertical bed thickness (V) and inserted into the above equation. Stratum contours must be constructed for the base and for the top of the same bed. For example, in Figure 7.2, stratum contours for the base of E and for the EF contact (top of E) may be compared. They are constructed across the map and then we estimate which stratum contour for the top of the bed directly coincides with the stratum contour for the base of the bed. For example, the contour for the base of E at 600 m directly coincides with the stratum contour for the top of E at 700 m. Thus, the vertical thickness $V = 100$ m. We may substitute this in the preceding equation to determine the more useful orthogonal thickness, t.

Note that in real situations and more realistic maps subsequently, stratum contours at round numbers of meters (e.g., 600, 700, etc.) for the base will not be directly overlain by round numbered top contours. One must then interpolate between the top contours to estimate (or calculate) the actual value of the top contour that would lie above the base contour. For example, if the TOP$_{700}$ contour lay halfway between the BASE$_{500}$ and the BASE$_{600}$ contour, we may surmise that the TOP$_{700}$ contour actually is superposed upon the BASE$_{550}$ contour. Thus, the vertical thickness $V = 150$ m. This may be substituted in the preceding formula to determine the orthogonal thickness t.

Method (3): True Thickness from Map Width of Stratum

When working with maps or aerial photographs, we may determine bed thickness t from the width of the outcrop of a bed, as long as topographic relief is relatively subdued. We need to know the dip angle and the outcrop width, w, measured perpendicular to strike. Figure 7.3(b) illustrates the principle. The trigonometric relationship we require is

$$\sin\,(dip) = \frac{t}{w} \text{ so that } t = w\sin\,(dip).$$

This method is useful on large area maps of gently dipping strata. Once topography becomes a consideration, this technique may be unreliable.

Figure 7.4 shows the relatively simple geology of Gotland, Sweden. Topographic relief is negligible. Calculate the thicknesses of the strata and draw a cross-section AB; you may assume there is no topographic relief and you may exaggerate the vertical scale so that the dips will appear larger. (For example, exaggerate the dips to 20°.)

Q.1. Is there any evidence for an unconformity in the sequence? Indicate this on the map and stratigraphic column.

Q.2. Which formations vary most in thickness? What are possible reasons for this?

Q.3. Identify the position of an outlier, probably due to a topographic high or minor dip variation. Which formation occupies the outlier.

Q.4. Estimate the approximate thickness of the Wenlock, Klinteberg, and Hemse formations from outcrop widths and the known dips. List the thicknesses against the stratigraphic column, with estimates of your precision (i.e., ± ? meters).

The next diagrams concern the Pre-Cambrian Shield beneath Manitoba (Figures 7.5 and 7.6). The topographic relief on this large-scale map is negligible; you may regard the mapped surface as a flat and horizontal plane. The Archean bedrock of the Canadian Shield in NW Ontario dips beneath the Paleozoic and Mesozoic sedimentary rocks on the Prairies of Manitoba and the dips may be inferred from Figure 7.6 which contours the surface of the Canadian Shield. In other words, the Archean Shield gently arches about an NW–SE axis running through Thompson. The sedimentary rocks dip very gently and it may be best to report them as gradient ratios rather than angles (use Table 6.1). Depths to the base of the sedimentary sequence (the top of the basement surface) are contoured in meters on this map. The depths are accurately known from boreholes and seismic surveys and well logs. (The area hosts oil and potash deposits that were the focus of geophysical exploration.) The contours shown are thus structure contours for the basement surface, which is also the unconformity at the base of the Paleozoic sedimentary sequence.

Q.1. Determine the slope of the basement beneath Manitoba at points A, B, X, Y, and Z (Figure 7.6) using adjacent structure contours. Be careful to measure separations in a perpendicular direction between contours. Indicate your answers on the

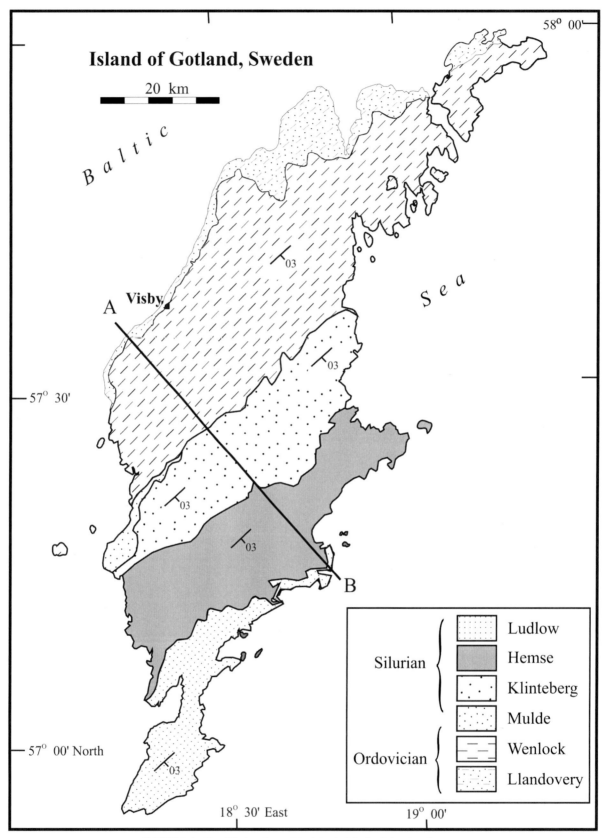

Island of Gotland, Sweden

20 km

Baltic

Sea

Visby

A

B

58° 00'

57° 30'

57° 00' North

18° 30' East   19° 00'

	Ludlow
Silurian {	Hemse
	Klinteberg
	Mulde
Ordovician {	Wenlock
	Llandovery

FIGURE 7.4 Simplified geology of the island of Gotland, Sweden. See text.

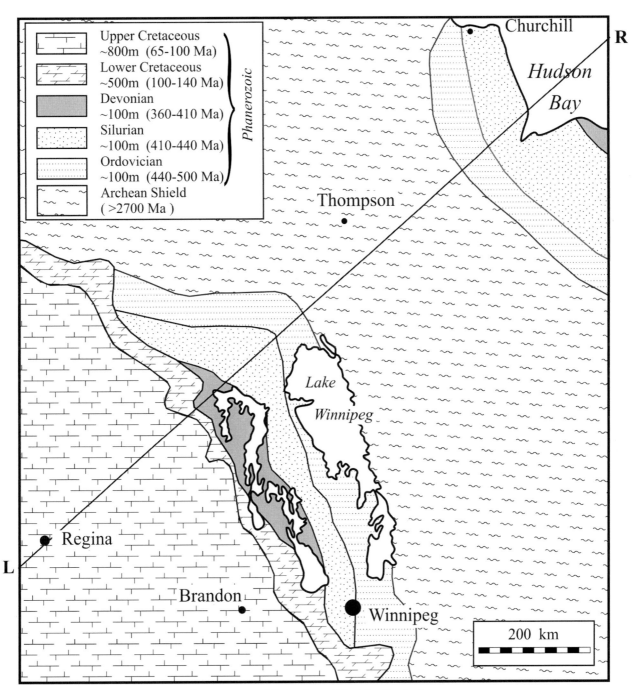

FIGURE 7.5 Simplified geology of Manitoba, Canada. See text.

map using the traditional dip-and-strike symbol, with the low-gradient angle expressed as a fraction (e.g., Table .6.1).

Q.2. Will the basal unconformity dip more or less steeply under Dakota and Saskatchewan than under Manitoba?

Q.3. Describe the feature represented by the closed loop contour at C.

Q.4. Draw a cross-section, left to right (LR), exaggerating the dips considerably (e.g., to about 15° for the Shield.) (What vertical exaggeration does your section represent?)

STRUCTURE CONTOURS FROM DIPS

Sometimes it is useful to predict the depth to certain bedding planes, in various parts of a map where that bedding

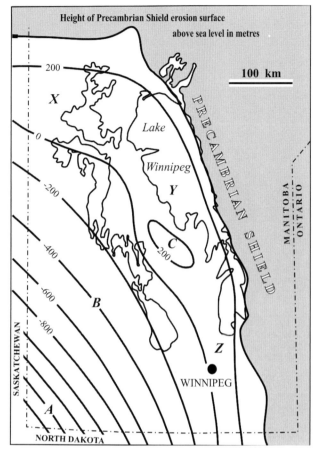

FIGURE 7.6 The Pre-Cambrian Shield beneath Manitoba, heights in meters above sea level. See text.

plane is not exposed. For example, it may be valuable to predict the underground course of an unconformity, which would also estimate the basement topography. If we examine Figure 7.7(a), we see that the Permian and Triassic sequences fill the Eden fault-bounded valley of northern England. The base of the Permian crops out along the western portion of the valley, west of the River Eden. At this point, we know the base of the Permian is at the surface so we can mark in a "zero–meter" structure contour along the boundary. We may ignore topography at this map scale. The trend of the contour is of course parallel to the strike shown by the dip-and-strike symbol. The dip angle is very approximate at 3°, and this corresponds to a gradient of approximately 1 in 20 (Table 6.1); this imprecision is not of great concern, the results will be approximate and that is sufficient in most exploratory geology. A bedding gradient of 1 in 20 implies that for every kilometer that we move across strike to the ENE, the base of the Permian should drop 50 m; alternatively, for each 5 km, the base drops 250 m. Thus, we may sketch estimated structure contours for the base of the Permian (Figure 7.7). Logically, these terminate against the boundary fault, and their trend or strike should be

concordant with the strike of the outcrop of the base of the Permian. Note that some unconformity surfaces are not planar but show relief; this is especially true if the sediments immediately following the unconformity are terrestrial. For example, where subaerial redbeds overlie a basement the unconformity surface may have hundreds of meters of relief. With such a sedimentary facies, it is very unwise to assume that its base is planar.

Q.1. Consider the Permian and Triassic rocks to the north of the area in the previous map (Figure 7.6(b)). Which sides of the faults marked "?" are downthrown? Mark the downthrown side with a tick.

Q.2. Extend the contours for the basement to the Permian from the Eden Valley to the north and to the west of Carlisle. Remember they should terminate against faults, they should follow the strike of the formations (shown by generalized dip-and-strike symbols) and you can infer their spacing from the dip angles shown.

Q.3. What can you conclude about the structure of the Eden Valley?

Q.4. How would you describe the structure of the area between the Solway Firth, Carlisle, and Wigton?

Figure 7.8 is a tectonic map of Western Canada; it shows areas of similar age of tectonic deformation. It does not primarily show ages of rocks. The Canadian Shield of Archean metamorphic rock dips gently westward beneath Phanerozoic strata. Different tectonic terranes lie further west. Note how the structure contours for the top of the Shield steepen and descend to great depths in the west beneath the Rocky Mountains. Using gradients rather than angles, mark the dips at various locations from the Arctic Ocean to Calgary and inland toward Saskatchewan.

Q.1. Why do the structure contours steepen in the west?

Q.2. SE of Calgary the structure contours for the top of the Shield form a closed loop, why?

Q.3. Indicate the ratios of the dip angles at several points from Saskatchewan to Calgary.

Figure 7.9 maps structure contours for the top of a Carboniferous petroleum-bearing horizon, the Eakring oil field, UK. In the previous exercises, we saw that in Nature, structure contours may not be straight but will follow the course of a horizontal surface in the geology, just like a topographic contour follows horizons on the hillsides. This example shows curved stratum contours for the top of a petroleum-bearing horizon. This indicates the strata were gently folded.

Q.1. Locate domes of the strata. These represent oil traps.

Q.2. Oil is present to a depth of 660 m. Shade in potential reservoirs defined by this basal level.

Q.3. Locate anticlines and synclines using traditional symbols.

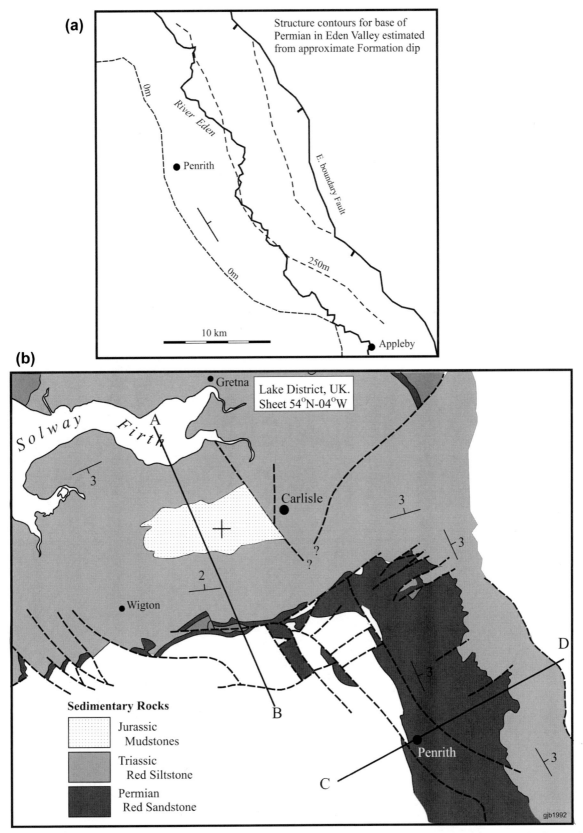

FIGURE 7.7 Simplified geology around Penrith and the Eden Valley, UK. (a) Structure contours for base o Permian, (b) geology with faults as broken lines. Draw sections along AB and CD.

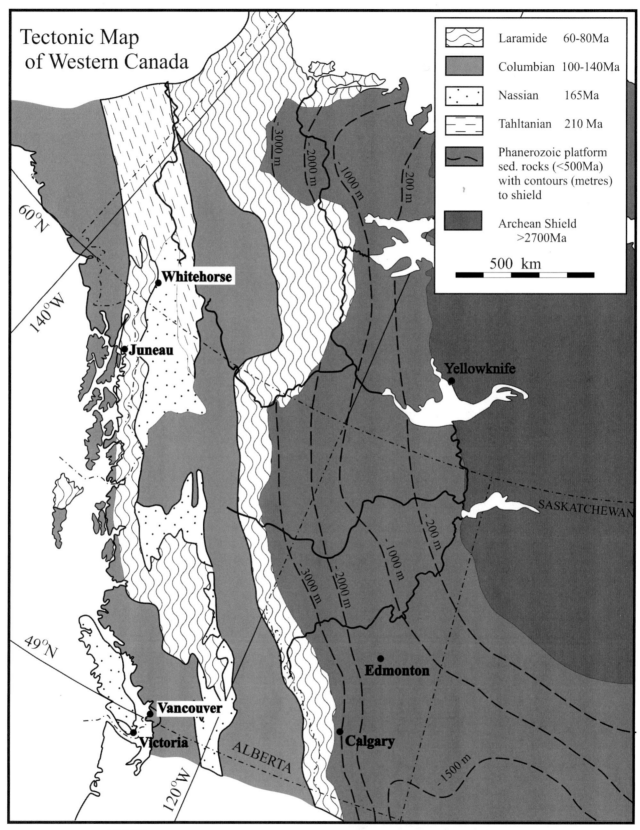

FIGURE 7.8 Tectonic map of Western Canada with contours on the base of the Pre-Cambrian Shield.

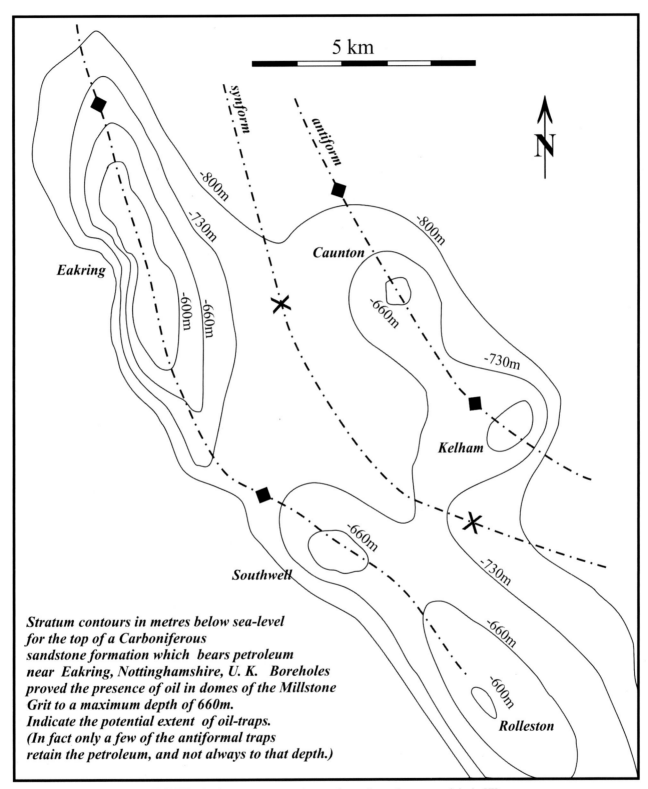

5 km

synform

antiform

N

-800m

-730m

Eakring

-600m

-660m

Caunton

-800m

-660m

-730m

Kelham

-660m

Southwell

-730m

-660m

-600m

Rolleston

Stratum contours in metres below sea-level
for the top of a Carboniferous
sandstone formation which bears petroleum
near Eakring, Nottinghamshire, U. K. Boreholes
proved the presence of oil in domes of the Millstone
Grit to a maximum depth of 660m.
Indicate the potential extent of oil-traps.
(In fact only a few of the antiformal traps
retain the petroleum, and not always to that depth.)

FIGURE 7.9 Stratum contours at the top of a small petroleum reservoir in the UK.

Unconformities

Chapter Outline

A complete stratigraphic record is to found nowhere; the rocks of the Earth's crust are constantly recycled by uplift, erosion, deposition, burial, metamorphism, melting, and crystallization. Thus, each location provides a fragmentary record of the stratigraphic column. Moreover, no "original" rocks survive in the Earth's crust. The oldest continental rocks are slightly older than ~4 Ga (billion years = 1000 Ma) and are at least a billion years younger than any solid rocks that formed at the surface of the Earth. Most of the continental crust is less than a billion years old; the only significant exceptions are the Pre-Cambrian Shields, all older than 600 Ma and almost exclusively older than 1000 Ma. To simplify matters, we may say that the continental crust is recycled by accreting sedimentary units at the margins of continents and by their subsequent erosion and redeposition in the ocean. Thus, the area of continental rock has grown through geological time. While some old parts have largely escaped some recycling, e.g., the Pre-Cambrian Shields, other rock units have had a very short residence on the continent. Mean residence times of the "average continental rock" are extraordinarily difficult to estimate but given the relative areas and ages of the continental crust, it may be of the order of 500 Ma. In contrast, the oceanic crust is recycled very quickly and much more effectively; it is created at mid-ocean ridges (MOR) and consumed at subduction zones on active continental margins (Figure 9.11). Since a typical ocean crust spreading rate is ~50 km/Ma from the MOR toward the active margin, the maximum half-width of the largest large ocean basin (~6000 km) implies that no intact ocean crust older than 150 Ma should survive. Accuracy of the calculations hinges of precise spreading rates and MOR-continent distances but is validated by geochronology and paleomagnetism of the ocean basins. The oldest ocean floors are found adjacent to the subduction zone, just before they are consumed, and their ages are ≤180 Ma in the Atlantic and Pacific Oceans.

We focus attention on continental rocks, which are much more varied in rock types and mineralogy than the ocean crust. The latter is composed primarily of basalt overlain by as much as 2 km cover of pelagic sediment. The more variable continental rocks may be plutonic, igneous, metamorphic, or sedimentary. Sedimentary and metamorphic rocks encompass most of the continental area but metamorphic, igneous, and plutonic rock comprises most of its volume since they dominate at depth. Ninety-nine percent of the continental crust comprises O, Si, Al, Fe, Mg, Mn, Ca, Na, K, Ti, and H and, due to a fundamental law of physical chemistry known as Gibb's phase rule, the number of minerals created in any igneous, metamorphic, or plutonic rock is usually ≤6. The higher grade the metamorphic or plutonic rock, the fewer the minerals. The most common rock-forming minerals (some of which encompass a considerable variation in chemical composition and petrological significance) include feldspar, quartz, amphibole, pyroxene, olivine mica, chlorite, and clay minerals. These silicates all use Si–O units as a fundamental-building block for their crystal structure; the variety of silicates being due to the arrangements of the Si–O units (tectosilicate, sheet silicate, chain silicate, etc.) and also due to the cations, principally Al, Fe, Mg, Mn, Ca, Na, and K, which fit in suitable spaces and must also satisfy electrical neutrality.

After silicates, the most common other sedimentary rock-forming mineral is calcite, which forms limestone. Unfortunately, the fascinating and important economic deposits and ore minerals are insignificant in terms of

Understanding Geology Through Maps. http://dx.doi.org/10.1016/B978-0-12-800866-9.00008-9

crustal volume (e.g., magnetite and hematite iron formations, evaporite deposits, chalcopyrite, pentlandite, pyrrhotite, and pyrite). The only minerals in this group that may be considered as marginally "rock forming" are the ubiquitous accessory minerals magnetite and titanomagnetite. In some cases, they form as much as 1% by volume of a crustal rock.

Igneous, metamorphic, or sedimentary rocks erode at the surface to produce new sedimentary material, which is deposited in strata that are initially mostly perfectly horizontal. In the strictest sense, "continuous sedimentation" is ephemeral, enduring for hours, days, months, or in rarely longer for clastic sediments. Chemical sedimentation, i.e., precipitation from supersaturation of some salt or carbonate, may be more persistent but this is arguable. The duration of continuous deposition depends on the sedimentary environment but eventually each interval of sedimentation ceases or pauses before it resumes or is succeeded by a different kind of sedimentation. A bedding plane marks the pause it is a discrete surface of no thickness, representing an unknown interval of time. Units of similar sediment are beds, separated by bedding planes. Thick strata without obvious bedding planes are described as massive; the inference is that depositional conditions were uniform or fast or both.

What time interval does a bed or a bedding plane represent? For intertidal sediment, successive bedding planes may represent about 11 h, the usual tidal frequency in the modern oceans. In lakes near the margin of a glacier, rhythmic layered silts (varves) mark each annual melting season; every 11th unit may be thicker, witnessing the 11-year cycle of enhanced solar activity. In the deep oceans, on average 4 km deep and beyond the continental shelf, deposition is virtually restricted to settlement of very fine material from suspension or precipitation from solution. Consequently, such pelagic sediments typically accumulate by a few millimeters per thousand years; each bedding plane might represent the absence or erosion of thousands of years' worth of sediment. Microfossils, especially foraminifera, and ^{14}C chronology of deep ocean drill cores have established these rates. Clearly, the time significance of bedding planes within a formation of similar sediment strongly depends upon its depositional environment.

Depositional environments and sources of sediment eventually change with the passage of time. Causes include, but are not restricted to changes in sea level, changes in the level or slope of the adjacent land (mountain building and other tectonic events), volcanic activity, changes in input from major rivers, and changes in submarine currents (especially due to plate tectonic rearrangements). Some depositional environments are quite ephemeral; for example, most river systems are very dynamic, changing their locations and conditions within a few millennia and over a few kilometers. On the other hand, some limestone/chalk sequences represent similar conditions over tens of millions of years and over huge areas.

Strata representing similar material and similar depositional environment gather into larger sedimentary units such as formations for which upper and lower boundaries are sometimes formally fixed in time and space. (Important examples such as those between geological periods require the intervention of an International Commission and may result in the placement of permanent markers in the field to locate important boundaries.) The smallest unit is the bed, which is assembled successively into members, formations, subgroups, groups, and supergroups. Most of the maps. we deal with in this book at a scale of 1:10,000 or 1:50,000 would show the distribution of beds, members, and formations. In some cases, these terms are less formally used.

Time gaps represented by nondeposition may occur within a formation but the geologist's experience allows him to recognize that they are insignificant compared to the time gaps between different, successive formations. The erosional surface between formations may represent a huge time interval and may be marked by a basal conglomerate. Clearly, without geochronology or paleontology, the importance of time intervals between formations is somewhat arbitrary and perhaps subjective. For example, formations within the Quaternary (6.5 Ma to present) are defined on the basis of much less fundamental stratigraphic breaks than for formations in the Jurassic, which encompassed 65 Ma worth of stratigraphic accumulation from 206 to 142 Ma (Figure 1.2). For the purposes of stratigraphic correlation between disconnected regions, and for geological mapping, the magnitude of the time difference between stratigraphic units is irrelevant.

The criterion for defining the significant bedding plane that separates strata of very different ages was recognized over 200 years ago. In about 1788, James Hutton made the critical and genial observation that if one formation overlay a different and previously tilted formation, the overlying formation must be significantly younger. The older sediment had to be lithified, tilted by earth movements, affected perhaps a mountain-building event (orogeny), then uplifted and exposed at the surface, all of which requires considerable time. Only then was an erosional surface available for the deposition of the second formation. Any significant break such as that recognized by Hutton is termed an unconformity. In other languages, "unconformity" is usually translated by a word equivalent to the English word "discordance". Discordance actually conveys the concept more meaningfully and is a better umbrella under which to describe the several types of unconformity (in English; disconformity, nonconformity, unconformity) that the geologist must understand.

Let us examine the notebook sketch of Hutton's unconformity (Figure 8.1). The outcrop is at Siccar Point, on the coast of Berwickshire, southern Scotland a couple of days by horse, south of Edinburgh. In 1788, Hutton recognized that the tilting, uplifting, and erosion of the lower formation

Notebook sketch of 3 m wide outcrop:

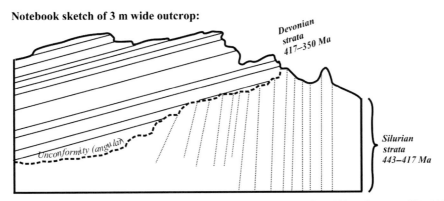

FIGURE 8.1 Notebook sketch of the first unconformity recognized by James Hutton, shortly after 1788 on the coast of Berwickshire, UK. The lower strata are Ordovician and the upper strata are Devonian. At least 70 Ma separates the two sets of strata.

took considerable time. Thus, a considerable time interval was required between the deposition of the old formation and the deposition of the young one lying upon it. Hutton could not possibly imagine the length of time represented by the unconformity; at that time, the Earth was believed to be just a few thousand year's old. His interpretation, although objective and sound, was sufficiently disturbing to contemporary theological thought that he suffered personally and professionally despite the relatively open-minded intellectual era known as the period of Scottish Enlightenment. Hutton was a medical doctor and professor at the University of Edinburgh. From paleontology, we know now that the lower strata are Ordovician and that the overlying beds are Devonian. From geochronological studies of volcanic ash beds and "bracketing intrusion ages" (Chapter 3), it is known that the Ordovician was deposited before 495 Ma and that the Devonian was deposited sometime after 417 Ma. Thus, the interval between the deposition of the two sets of strata is ≥78 Ma. The type of unconformity or stratigraphic discordance that Hutton recognized is very convincing due to the difference in dips of the beds above and below the unconformity. Moreover, the lower beds are also metamorphosed, bearing a slaty cleavage due to orogenic movements. Because of the contrasting dips, this is specifically known as an angular unconformity. Where it defines the base of an important overlying formation, it may also be recognized by the presence of a basal conglomerate (e.g., see basal conglomerate to Carboniferous in the English Lake District, Figure 6.7).

It is important to remember that an angular unconformity requires a difference in dip or strike of the underlying and overlying strata. Thus, beds below an angular unconformity and above may have the same dip angle or the same strike. However, the dip or the strike or both must differ. This depends on the cross-sectional view of the geology (cut affect and real/apparent dips). This problem is confounded by the angle at which the outcrop is viewed (parallax) since the outcrop is actually a rough three-dimensional surface. (This problem may be reduced if it is possible to view or

photograph the outcrop in the same direction, from a greater distance.) It is more instructive to make a notebook sketch that simplifies the actual outcrop to essential geological relationships in cartoon form as if they were projected onto a plane surface. The effort of making a scaled sketch helps one to understand the relationships; quickly snapped photographs are often quite difficult interpret at a later date. The pioneering structural geologist Gilbert Wilson (who taught John Ramsay) believed that one could not understand an outcrop fully unless its essence was extracted and summarized in a sketch.

TYPES OF STRATIGRAPHIC DISCORDANCE

Angular unconformity older strata are tilted, eroded, and exposed as a depositional surface and an unconformable sequence of younger strata are deposited above them. There is a difference in strike or dip or both between the two sets of strata. (Note the dip angles could be the same, but at least the strikes would be different.) A significant angular unconformity may have considerable relief of the erosion surface or a basal conglomerate upon it or both.

Disconformity strata are eroded to yield a new depositional surface. The unconformable overlying strata have the same strike and dip as the underlying strata. This may be identified by different fossils or by examination of the actual contact in the field. It is not evident from a map, unless the legend indicates the two sets of strata are of very different ages, e.g., if they were Eocene and Pliocene, the entire Miocene sequence was eroded.

Nonsequence. As with a disconformity, the younger and older strata are conformable; they have the same strike and dip. However, there was no erosion; some beds were simply never deposited.

Nonconformity. Where a sedimentary formation overlies an igneous or metamorphic rock there is obviously a significant stratigraphic break. However, since there was never a possibility of any kind of conformity, it is distinguished by this term.

Syndepositional discordance: In most elementary introductions, emphasis is placed on unconformities of tectonic significance. Earth movements tilted and possibly folded earlier lithified rock so that subsequent, much younger strata were unconformable. However, angular unconformities representing short periods of erosion are common in fluvial systems. As river channels meander, they cut into previous river deposits laying down layers that are unconformable on the earlier ones. No earth movements are involved; no beds are tilted. The angular differences between the beds arise because they are deposited on channel banks and slopes; younger slopes strike and dip differently from older ones giving rise to the unconformable relationship. Such syndepositional unconformities cause thickness variations in sands and gravels. These variations may be determined from contours of thickness (isopachytes) that we may discuss later. Submarine erosion is documented in several parts of the stratigraphic column. For example, in the Western Mediterranean, the Cretaceous–Paleogene limestone deposition was considered to have been continuous until the absence of ~15 Ma worth of sedimentary rock was identified. This is quite subtle in the field; it is not marked by an angular unconformity or basal conglomerate.

STRATIGRAPHIC DISCORDANCE IMPLIED BY DIFFERENCES IN DEGREE OF REGIONAL METAMORPHISM

Angular unconformity is the most convincing visual evidence for a major stratigraphic break, usually one that postdates a major phase of tectonic deformation, perhaps even mountain building (orogeny). In elementary studies of geology like this one, instructors commonly avoid introducing the notion of metamorphic discordances. This is because metamorphism is undoubtedly the most difficult aspect of petrology in undergraduate study; it demands good petrographic, mineralogical, and geochemical skills combined with knowledge of the petrology of the protoliths (i.e., igneous or sedimentary rock from which the metamorphic rock is derived). Moreover, in the last 2 decades, it has become apparent that the textures of metamorphic rock cannot be divorced from all the other aspects, thus, structural petrology and materials science are essential to the understanding of the solid state processes by which metamorphic rocks develop. Sadly, metamorphism is not very enthusiastically embraced by the undergraduate curriculum.

At this stage however, we can use some basic metamorphic ideas to recognize major discontinuities between rocks. This discussion only applies to regionally metamorphosed rocks, those which form accompanying the formation of mountain belts or greenstone belts in the Archean. Such rocks are classified by the range of temperature and pressure under which they reach equilibrium. "Grade" of metamorphism is a qualitative measure of temperature.

Since grade is characterized by an index mineral it is obviously dependent on the availability of a suitable rock in which that mineral may form. For example, in metamorphosed mud rocks, we may see the following sequence of key or "index" minerals with advancing temperature/

Mud rock	Clay minerals
Phyllite	White mica, sericite
Slate	Muscovite, chlorite
Schist	Garnet, muscovite, amphibole
Granulite	Garnet, pyroxene, feldspar

Differences in grade may be mapped in the field to locate surfaces termed isograds. In reality, the subject is much more precise and detailed than this.

A more useful classification of metamorphic rocks that associates with a combination of pressure-and-temperature conditions is metamorphic facies. A metamorphic facies are represented by a broad range of rock types and chemical compositions but certain ranges of pressure and ranges of temperature always produce the same association of a few key minerals. Facies and grade are quite distinct concepts and cannot be used interchangeably. In general, grades can be mapped out on the basis of key minerals; facies cannot since they represent associations of lithologies at similar metamorphic conditions.

In this book, we cannot delve further into these matters except to note that certain metamorphic rocks could not associated with one another in a concordant stratigraphic sequence. A major tectonic contact or an unconformity would be required to separate the two kinds of metamorphic rock. If the contact was an unconformity, the lower grade metamorphic rocks would normally be the youngest. Table 8.1 presents a very simple idea of a few metamorphic rocks types of contrasted high and

TABLE 8.1 A few Important Metamorphic Rocks (a–c) that would be Discordant with Adjacent Rocks

(a) Fault rocks	(b) High grade	(c) Low grade
Mylonite	Granulite	Slate
Cataclasite	Charnockite	Phyllite
Pseudotachylyte	Eclogite	Quartzite
	Plutonic granitoid	Marble (most)
	Migmatite	Greenschist
	Gneiss	Greenstone
	Anatectite	Blueschist
	Skarn	Brownschist
		Epidosite

low metamorphic grade; they would not be expected in conformable contact. Metamorphic fault rocks are also included in a separate column. In nature, there are of course transitional lithologies that may exist to form a conformable sequence between those in the middle and left columns. However, that does not concern us here, we are interested only in the possibility of recognizing discordant stratigraphy from incompatible metamorphic neighbors. After studying metamorphism, one learns that there are many more subtle incompatible metamorphic rocks whose juxtaposition would require some discontinuity, e.g., a fault or unconformity.

ONLAP AND OFFLAP WITH ANGULAR UNCONFORMITIES

On a map scale, it is possible to recognize two subtleties of angular unconformities that are important for the understanding of the evolution of depositional basins (Figure 8.2). Clearly, angular unconformities may accompany the crustal sagging associated with significant sedimentation. However, it is also possible to determine fluctuations in the advance or retreat of coastlines from the relative extent of the strata overlying the unconformity. This topic is introduced here more for completeness and assistance in further studies.

Angular unconformity with onlap. The overlying, unconformable sequence may be deposited while the land is sinking or the sea is rising. In either case, successive strata encroach further on the depositional surface. Map distributions and constructions using structure contours reveal this relatively easily.

Angular conformity with offlap. In this case, the depositional basin is shrinking so that successive strata are deposited over smaller areas. This is more difficult to confirm from a map since it may be confused with subsequent erosion of the younger sequence.

NONSEDIMENTARY CONTACTS

Most tectonic contacts occur where rapid earth movements due to earthquakes may have caused faults or joints. Joints have very small displacements so their effects are not evident from a map (recall Figure 4.3). Most faults are large enough to represent on some scale of map; they may extend for tens of meters to hundreds of kilometers across the map. Typically, the displacement or total motion on the fault is a few percent of its outcrop length. The fault length may be meters to hundreds of kilometers.

As we shall see later, the sense of motion on a fault is not immediately apparent from the displacement of adjacent geological boundaries; this subject requires further discussion later but the problem was introduced in connection with the map of the English Lake District (Figure 6.7). Very large displacements occur only where the motion is lateral, as with a wrench fault, or a very special fault, initiated in the ocean floor, called a transform fault, which may have enormous displacements. Tear and transform faults are normally vertical fractures and appear therefore as straight lines across the map. Some faults move intermittently over very long periods of time so that older strata are more displaced than younger strata; these may be termed growth faults.

Other tectonic contacts may occur due to slow incremental distortion (strain), they are not necessarily brittle fractures but may be at least partly ductile in character. They include thrusts, slides, and shear zones.

Intrusive contacts. These are mostly contacts between intrusive igneous rocks and the rocks that they intrude. In the case of a sill, the contacts are conformable; the sill is parallel to stratification or sometimes a metamorphic structure like schistosity. A sill is not necessarily horizontal, the definition is based on conformity with an adjacent structure, normally bedding. A laccolith is essentially a sill that bulges upward but usually rather large. A dike is a disjunctive sheet; it cuts across strata, prominent layering and perhaps across other structures. It is defined by its crosscutting relationship, not by its orientation. However, for mechanical reasons, many dikes are vertical sheets. A pipe, e.g., volcanic pipe or kimberlite pipe, is a nearly cylindrical vertical body of igneous rock.

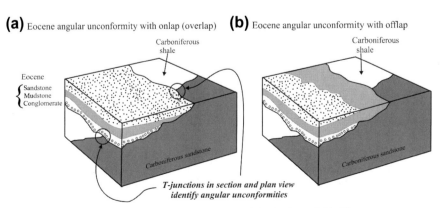

(a) Eocene angular unconformity with onlap (overlap)   **(b)** Eocene angular unconformity with offlap

Carboniferous shale

Eocene
{ Sandstone
  Mudstone
  Conglomerate }

Carboniferous sandstone

Carboniferous shale

Carboniferous sandstone

T-junctions in section and plan view identify angular unconformities

FIGURE 8.2   Angular unconformities with (a) onlap and (b) offlap.

For most igneous bodies, especially those with rounded forms in map view, it is very difficult to predict the orientations of the contacts underground. This is because the contacts are not planar; geophysical exploration methods such as magnetic, gravity, or seismic surveying are required. Large bodies of igneous rock may be termed plutons or batholiths; some batholiths (e.g., Peru) are larger than small countries. The irregular and unpredictable orientation of the contacts of plutons and batholiths is attributable to their modes of emplacement into the country rock. These are very complicated matters, to which the following does not do justice. However, the following terms and their meaning may help you to appreciate why the contact orientations are unpredictable.

Stoping. This describes the passive sinking of country rock into the magma chamber so that the magma chills against the fractured country rock. The igneous rock then takes up the shape of the voids (sometimes called cauldrons or caldera) in the country rock.

Anatexis. Melting of the country rock [usually partial melting] to form the magma. The melting front is naturally highly irregular, dictated by temperature, fluid movement, and the melting point of the country rocks at different locations.

Metasomatism. Diffusive transformation of country rock to one of igneous appearance and mineralogy typical of an igneous rock. The transformation is largely in the solid state by chemical diffusion of cations along suitable chemical potential gradients. Such high-temperature diffusive processes are not really distinct from high-temperature metamorphism and the rocks are sometimes termed "plutonic" rather than "igneous". Most "granite" of the Canadian Shield is not igneous (=crystallized from form magma) but plutonic granitoid, formed by metamorphic diffusion which transforms already metamorphosed sedimentary and igneous rocks into homogeneous crystalline rock that superficially resembles an igneous rock. This notion is nicely conjured up by the old but disfavored term "granitization".

Diapirism. This is the forceful emplacement of rock caused by its lower density or gas/vapor pressure. The intrusion bubbles up through the country rock bending it aside. Certain plutonic–metamorphic rocks with granitic–granodioritic composition also emplace diapirically; excellent examples are known, mainly from the Archean, especially in northern Ontario. One group of sedimentary rocks, the evaporites (=salt strata) may also form diapers. Due to their low density, evaporite horizons may float upward (very slowly in the solid state) as bubbles though the overlying strata to form salt domes. Classic salt domes are found in Texas, California, and Northern Germany.

Return to the map of geology between Manitoba and Hudson's Bay (Figure 7.5) in Chapter 7. Detect unconformities and review.

Q.1. Using stratigraphic information and the nature of the mapped boundaries ("T-junctions") indicate the presence of any unconformity, disconformity, or nonconformity in this region. (Mark it on the stratigraphic column and on the map boundary using a colored or wavy line.)

Q.2. From the stratigraphic order place dip-and-strike symbols (no dip angle is required) at locations in the Phanerozoic outcrops, close to the line AB across the map.

Q.3. Using the approximate formation thicknesses, given in the stratigraphic table, estimate the angles of dip for the Ordovician, Silurian, and Lower Cretaceous formations SW of Lake Winnipeg. Since the angles are very small, it is easier to work with the gradient ratios.

Q.4. Is there evidence for overlap during Phanerozoic deposition? Describe this.

Q.5. What is the most probable cause for the variation in dips of the Phanerozoic formations between Thompson and Regina?

UNCONFORMITIES AND SUBCROPS

This figure reveals the underground termination of a sandstone bed against an unconformity (Figure 8.3(a)). We shall see later that this subcrop or underground termination may be calculated and constructed on the map; it may prove a useful estimate of economic limitations on resource extraction. The underground termination of the sandstone is fixed by the intersection of structure contours for the bed and the unconformity; this will become clear later. Figure 8.3(b–d) shows how the unconformity (c) and then the sandstone bed may be back rotated to possible original orientations. This involves some assumptions that may be incorrect, for example, the untilting axes need not be horizontal. A subcrop termination will be determined in exercise (Figure 8.5).

UNCONFORMITIES AND STRUCTURE CONTOURS

Structure contours provide a deeper understanding of unconformities. Figure 8.4 shows a simple, hypothetical situation; two copies of the map are present for practice and the partial solution is shown in Figure 8.5.

1. First, identify the unconformity; in this case, the "T-junction" on the map is the obvious clue. (In other maps, the presence of a conglomerate horizon may or information about a large time gap in the stratigraphy would be equally useful). Mark the unconformity outcrop on the map using a colored pencil or a penciled wavy line. Also, place this in the stratigraphic legend

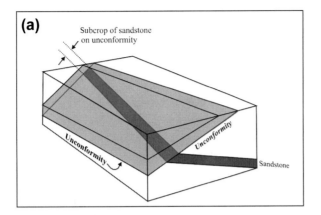

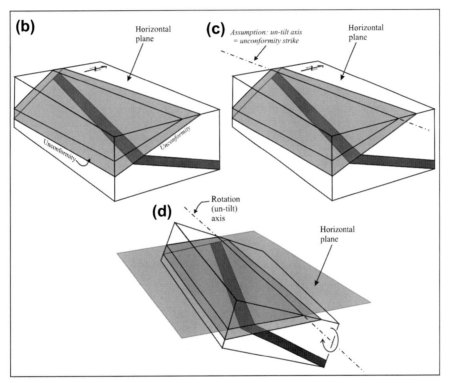

FIGURE 8.3   (a) Subcrop of a bed on an overlying unconformity. (b–d) Back rotating the strata in (a).

at the appropriate level. Next, establish the top and the bottom of each bed by imagining oneself walking uphill across bedding planes.

2. The next step is to construct structure contours for the unconformity surface. This is done by tracing the unconformity across the map and identifying where it intersects topographic contours. Identify each intersection with a small circle (just two examples are shown in Figure (8.4a) for the 300 m topographic contour). The two intersections of the unconformity with the 300 m topographic contour define the horizontal structure contour for the unconformity at 300 m. Draw this structure contour as a straight line and label it appropriately (e.g., u/c300). See Figure 8.5 for the solution.

3. Repeat the procedure where every topographic contour intersects the unconformity, carefully labeling the structure contours uniquely. You may extrapolate these contours by equal spaced parallel lines to parts of the map without topographic unconformity intersections. The complete construction is shown in Figure 8.5.

From the spacing of the unconformity's structure contours, you may determine the unconformity's dip from tangent (dip) = drop/separation (or use Table 6.2). Indicate the dip angle and strike on the map using the dip-and-strike symbol.

4. Now, we turn our attention to the beds underlying the unconformity (Figure 8.5(b)). In the example, the author chose to work with the top of the shelly limestone (structure contours labeled TSL300, etc. Intersections

(a)

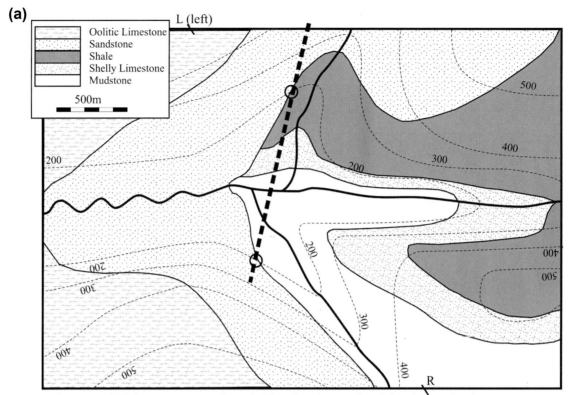

Construct and label structure contours for one bedding plane above, and one below the unconformity.
Indicate the dip and strike on the map using the usual symbol.
Construct a true-scale cross-section from A to B, with A on the left.
Show all constructions, in faint pencil, on the map & section.

(b)

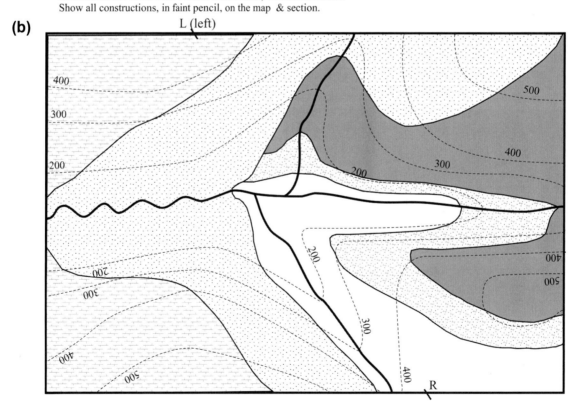

FIGURE 8.4   Two examples of an unconformity map to solve (see text). The initial structure contour for beds above the unconformity and below the unconformity are shown in (a) and (b), respectively.

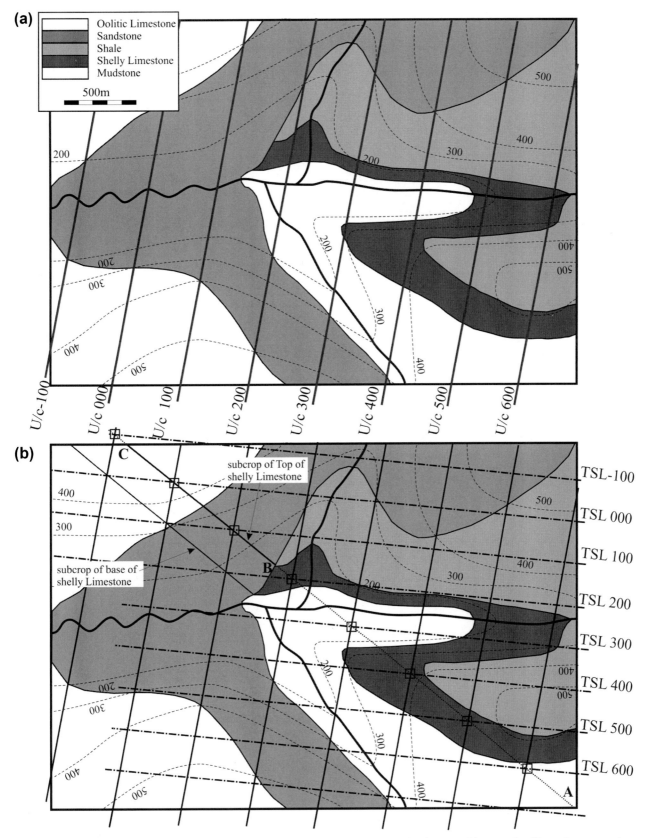

FIGURE 8.5 Construction of the subcrop of a shelly limestone horizon on the base of and unconformity. (a) partial construction (b) complete construction.

of structural contours for the top of the limestone and the unconformity (at the same elevation) fix the subcrop, which is drawn with a heavy line CB; this is the subterranean termination of the top of the limestone on the unconformity. Similarly, the subcrop for the base of the limestone may be constructed (Figure 8.5(b)). The subcrop defines the area underlain by limestone but it may be extrapolated SE across the map to show where the former limits of limestone where before erosion.

5. For completeness, as with all maps, we should determine the dip and strike of the strata and indicate this with the appropriate symbol. (You should always place this symbol very close to the outcrop of the bedding plane to which it refers.)

Do not be concerned if your structure contours are not precisely parallel or located exactly like those in the text. Small drafting errors do not affect the outcome of such a study. With real maps, "planar" beds are never perfectly planar as in these hypothetical maps so that structure contours are slightly irregular. Recall the Alston map with its natural and slightly irregular structure contours for the firestone sandstone (Figure 5.10).

In Figure 8.5(b), we "saw through" transparent overlying formations to observe the subcrop of a sandstone on the overlying unconformity. A subcrop is the pattern of beds on some buried surface, usually an unconformity. It may be economically important to understand the location of subcrops. For example, miners may need to follow a certain horizon beneath an unconformity but they need to know where that horizon terminates against unconformities. In addition, subcrops may the sites for reservoirs of valuable commodities such as water, petroleum, or natural gas. It is also important to remember that some unconformities are ancient weathering surfaces upon which economic minerals may have been enriched (e.g., gossans and placer deposits). Whereas Figure 8.5 permits us to visualize a subcrop, it is possible to determine its location beneath the mapped area with the techniques we have already learned.

PREUNCONFORMITY DIPS

This causes some concern in tectonics and in paleomagnetism. The question is far less easily answered than it is posed; "which way were the underlying strata oriented before they were tilted?" In other words, we have to restore the younger formation and the unconformity surface to horizontal. That same geometrical rotation should then bring the underlying strata to the orientation they possessed at the end of their original tilting or deformation phase.

At a very elementary level, we may imagine in our minds eye, the untilting of the unconformity develops an approximate idea of its modification of the underlying beds. Consider Figure 8.3; the observed geology may be back rotated until the overlying beds are horizontal. The

unconformity is now restored to its original horizontal dip and the underlying beds have their orientation shown after original tectonic tilting. If we assume the unconformity was simply tilted about its strike, we may geometrically reverse this procedure by untilting the unconformity to horizontal about its strike until the unconformity is horizontal.

In tectonics and paleomagnetic research, this restoration is done quite precisely but it does involve assumptions that are often overlooked. One problem is that the unconformity may not just have been rotated about its strike (assumed horizontal rotation axis). It is quite possible that there was also some rotation about another axis, e.g., a vertical axis due to an underlying thrust or due to plate rotation. The second problem arises where the rocks are not merely tilted but where they are squeezed and change in shape. This is formally termed "strain" in structural geology. Strain changes dips and strikes quite independently and perhaps accompanying physical rigid body rotation that we call tilting. It is extremely difficult to restore the dips of beds that have been affected in this way and requires advanced structural techniques. Paleomagnetic studies in tilted or strained rocks are plagued by these problems and sometimes throw plate tectonic reconstructions into question. Thus, you have already met techniques and ideas that are central to the largest problems in geology.

EXERCISES WITH UNCONFORMITIES

The legends that show the lithologies are not in stratigraphic order (Figures 8.6–8.9). Thus, you must first determine their stratigraphic sequence by "walking" uphill from a valley bottom, revealing the order of superposition of the beds.

Figure 8.6(a) shows the structure contours that you will need to use. Mark the unconformities on the map (you may identify them from "T"-junctions). Figure 8.6 shows some of the structure contours you will need and Figure 8.6(a) is a working copy for you to use. Determine the dip and strike of each set of beds and show it on the map using the appropriate symbol. Draw a cross-section AB with A on the left. Be careful to use the techniques of Figure 7.2 so that the appropriate apparent dips appear in the section.

Figure 8.7, identify multiple unconformities (every stratigraphic unit is unconformable). Determine the dip and strike of every nonmetamorphic formation and indicate it with an appropriate symbol. (None of the strata are horizontal.)

Figure 8.8 shows the geology of a small part of NW Scotland. From relationships to topographic contours, can you infer the dip and strike of the Moine schist and mafic schist SW of Tongue, on the southern border of the map? Similarly, determine the sense of dip and strike of the Cambro-Ordovician rocks; the area west of Loch Eriboll is easiest to show this. Mark the unconformities on the map. Note that the horizontally bedded Torridonian red sandstones were deposited on an area of very rugged relief; indicate places on the map where you can demonstrate this.

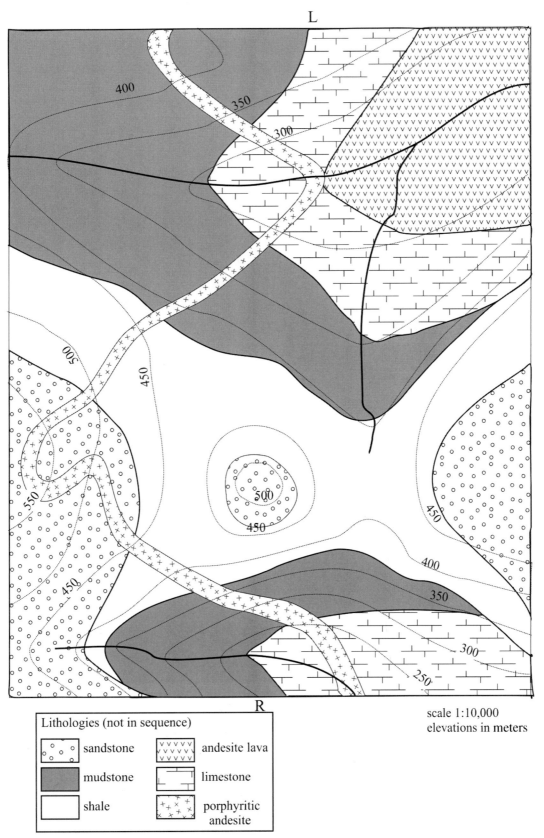

FIGURE 8.6   Simple map with two unconformities; see text.

FIGURE 8.7 Map showing the outcrop of a mineral vein, as a heavy dark line. This is partly concealed by the unconformably overlying limestone. The dashed line shows the subcrop of the mineral vein on the unconformity surface.

Figure 8.9(a) shows the small area around Golspie on the east coast of the Scottish Highlands. Locate unconformities and nonconformities in Figure 8.9(a) and estimate the sense of dip and strike wherever possible, indicating this on the map with an appropriate symbol. Determine the relative ages of the faults. In Figure 8.9(b), we see an extract of the main map. From the relationship between the horizontally bedded Devonian rocks and the topography what can you infer say about the topographic relief of the surface on which the Devonian rocks were deposited?

THICKNESS VARIATIONS: ISOPACHYTES

One consequence of an unconformity is that any angular discordance will produce thickness variations in underlying layers due to their partial erosion (Figures 8.10–8.13).

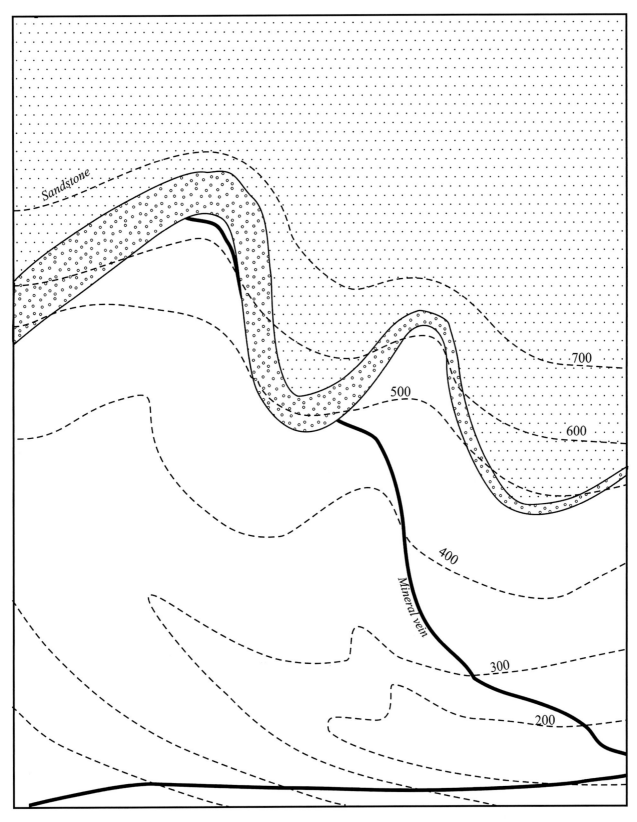

Elevations in meters   scale 1:10,000

FIGURE 8.8   Map the subcrop of the mineral vein beneath the unconformity.

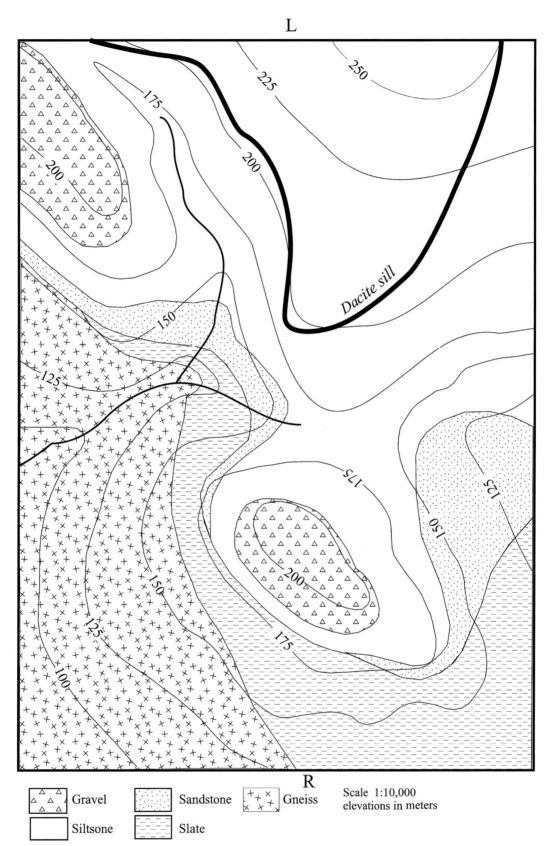

L

R

Dacite sill

Scale 1:10,000
elevations in meters

△ △ △ Gravel   Sandstone   Gneiss

Siltsone   Slate

FIGURE 8.9 Multiple unconformities; can you identify them all and mark the dip and strike of each formation?

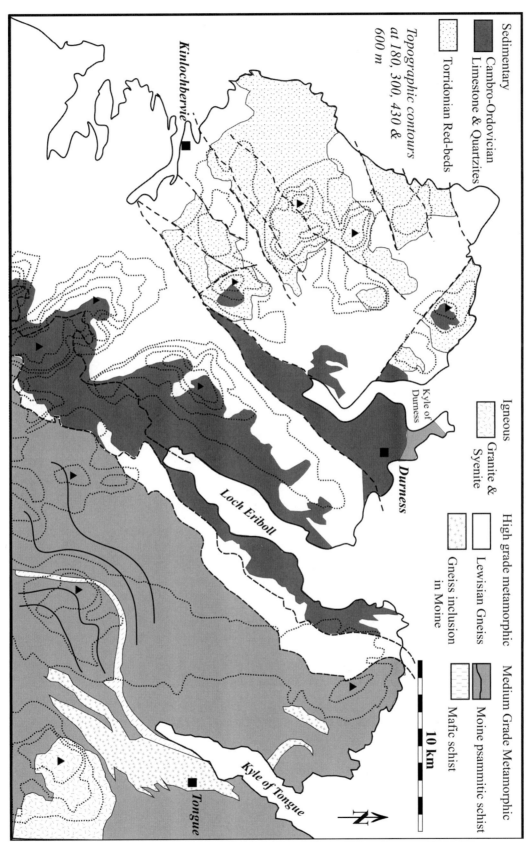

FIGURE 8.10   Unconformity and other contacts in NW Scotland.

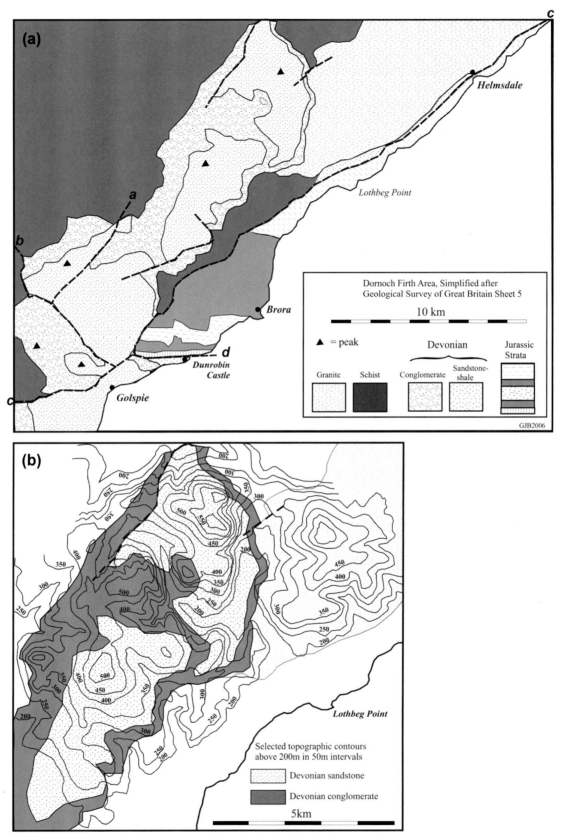

FIGURE 8.11 Unconformities and other contacts near Golspie, NE Scottish Highlands. (a) geology, (b) topography and simplified geology.

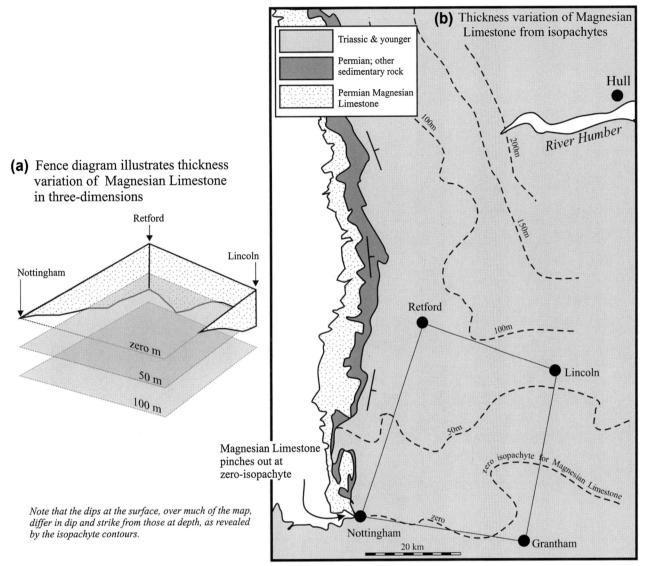

(a) Fence diagram illustrates thickness variation of Magnesian Limestone in three-dimensions

(b) Thickness variation of Magnesian Limestone from isopachytes

Magnesian Limestone pinches out at zero-isopachyte

Note that the dips at the surface, over much of the map, differ in dip and strike from those at depth, as revealed by the isopachyte contours.

FIGURE 8.12 (a) Fence diagram showing the thickness variations of (b) Magnesian limestone.

Of course, thickness variations may also be due to variations in depositional conditions. In either case, the technique discussed here may be applied to variations in thickness of any origin. For example, beds may be thicker in the center of a channel or depositional basin than at its edges. In some marine environments, sediment thins as it drapes over bathymetric highs and is thicker in depressions on the seafloor. The word isopachyte translates literally as "equal thickness" and the implication is that isopachytes drawn on maps will join points beneath which a certain bed has constant bed thickness. However, for simplicity, the methods of construction of isopachytes (Figures 8.14 and 8.15) usually fix the vertical thicknesses (*vt*) of the stratum of interest. Techniques usually involve determining the difference between the stratum contours for the top

and for the base of the bed or the use information from vertical bores.

Since isopachytes are more precisely contours of vertical thickness (*vt*) than of bed thickness (*t*), it is obvious that they are most easily interpreted and constructed where the strata have shallow dips. Where bed dip $= \theta$, the relationship between vertical and bed thickness is $t = vt \cdot \cos \theta$. As a general rule dips should not be greater than 20° since $\cos(20°) \approx 0.94$ causing true thickness to be overestimated by 6% which is probably the maximum error one could permit in geological applications. Of course, in serious investigations, it is easy to apply trigonometric corrections to improve the precision of the isopachytes. Bed thickness variations are important in determining the volume of petroleum or natural gas reservoirs, placer deposits, and

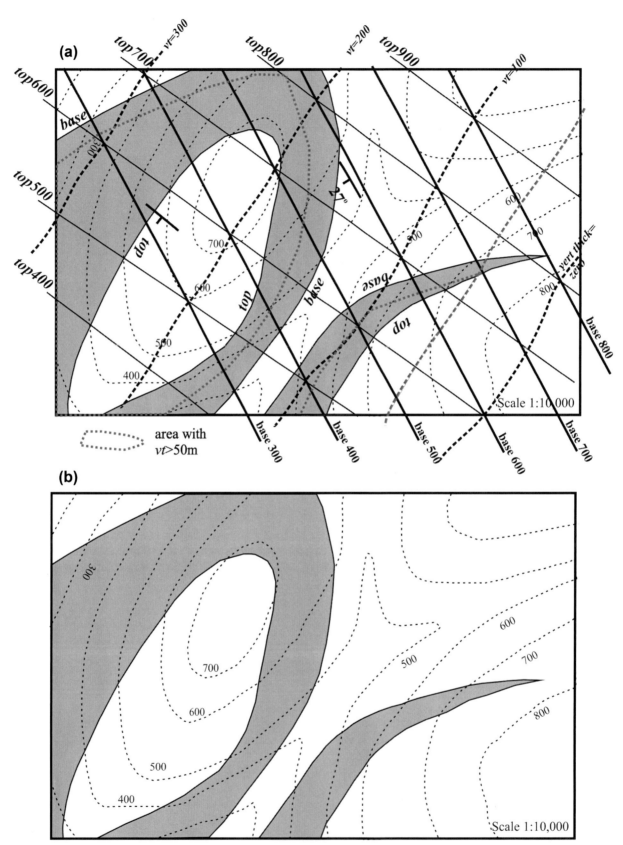

FIGURE 8.13 Partly completed map to show construction of isopachytes.

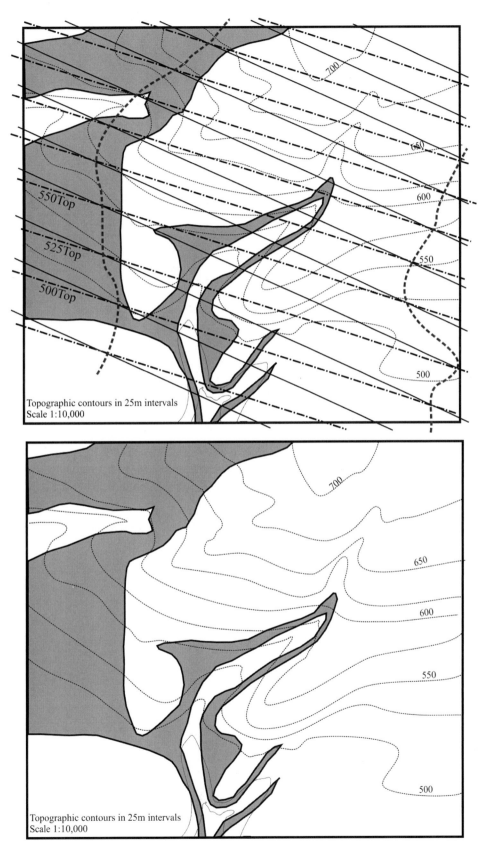

FIGURE 8.14 Further example of construction of isopachytes.

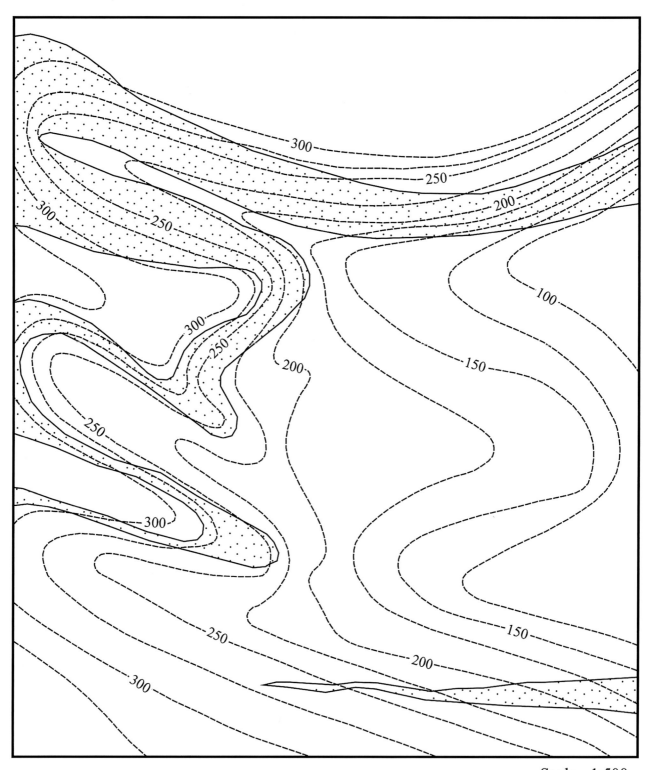

FIGURE 8.15 More complex example of isopachytes.

superficial deposits like sand and gravel used in construction. Fortunately, in most economic sedimentary deposits, dip angles are low. Figure 8.10 shows the regional variation in thickness of Magnesian limestone in eastern England. The map shows the mapped isopachytes, revealing that the limestone thickens from zero meters at Nottingham to over 200 m in the NE, near the River Humber. Figure 8.10(a) is a fence diagram that illustrates the thickness variation in three dimensions.

Figure 8.11 shows a worked example of an isopachyte map. First, the top and base of the bed must be determined by "walking out" a hill slope. Then, structure contours are constructed for the base and for the top of the bed. Their intersections determine the vertical thickness of the bed

and they may be connected to form contours of thickness or isopachytes. Figure 8.12 is another working copy since you may find it necessary to repeat this exercise in order to understand it.

A useful extension to this problem is to determine where there is an overburden limit on exploitation. For example, we may wish to exploit the oil shale only where more than 50 m thickness is present. To determine this (Figure 8.11(a)), we construct structure contours 50 m above the base contours; these intersect with topography to outline the area that has a vertical thickness (vt) > 50 m.

Figures 8.13 and 8.14 present further examples of an isopachyte problem. Construct the isopachytes; the top and base contours give a hint in the top diagram.

Faults

Chapter Outline

Whereas rocks are hard and brittle in our everyday experience, most of us have learned' indirectly that over long periods of time rocks have changed shape by flow or fracture. Some idea of was observed during the Renaissance, as shown by the sketches of folded and jointed sedimentary rock by Leonardo da Vinci. With our modern understanding of the extent of geological time and the slowness of Earth movements, it is easier for us to appreciate that rocks may change in shape or deform. Deformation is usually by some combination of fracture and flow. Fracture involves some physical separation of rock parts, usually along grain boundaries between mineral grains and the fracture has some regularity controlled by the adjacent stress trajectories. Flow results from rearrangement of grains along grain boundaries of other surfaces by particulate flow, by crystal plasticity or some combination. It includes many aspects of shape change described by terms such as extension, flattening, shearing, bending, and folding. Mostly, deformation is heterogeneous; i.e., straight lines do not remain straight and parallel lines do not remain parallel. Homogeneous strain is defined as the opposite and is largely a theoretical abstraction. In what follows, we shall set aside the complication that many natural rocks are anisotropic; some or all of their properties, including strength, have different values along different axes. Of course, some rocks do have similar properties in every direction; they are said to be isotropic. Compared to other chapters, this one has rather more theoretical introduction. The reason for this, as it will become clear, is that structures form from the application of stress but the intensity, duration, and constancy of stress determines whether faults or folds develop.

Clearly, relatively rapid earth movements favor fracture. Several physical conditions conspire to cause rocks to fracture rather than flow. First, the average surrounding pressure (confining pressure) must not be too great. Second, temperature must not be too great. Third, fluid pressures should not be too great.

Regarding pressure, the hydrostatic formula gives an acceptable approximation to the pressure caused by the weight of rock at depth (z) (Figure 9.1(a)). Given a column of rock of cross-sectional area A, its volume above depth z will be Az and if the density of the rock is ϱ, its mass will be ϱAz. The force (=mass × acceleration) at the base of the column of rock will therefore be ϱAzg, where g, the acceleration due to gravity, may be regarded as constant worldwide (and to the base of the mantle) for the precision required in this sort of calculation. Near the surface, especially where there is rugged relief, shallow stress trajectories will veer from the deeper equivalents (Figure 9.2).

Stress or pressure is force divided by the area over which it acts. Therefore, the vertical pressure at the base

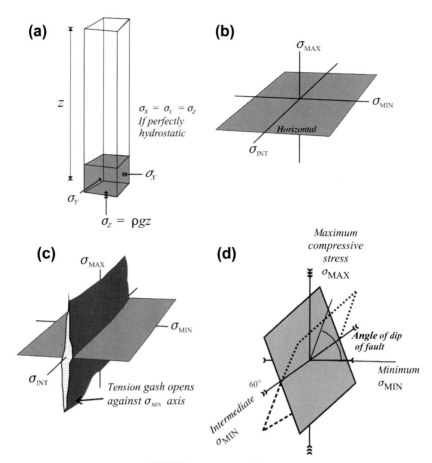

FIGURE 9.1   Stress definitions.

of the column is $\varrho Azg/A = \varrho zg$. Density of the rock column, ϱ is constant so we write "constants" first to give vertical pressure or stress $= \varrho gz$. We must remember to work with consistent SI units throughout (m, kg, s). Acceleration due to gravity may be taken as ~9.8 m/s² (more precisely, it varies several percent depending on latitude and altitude). The density of consolidated rocks varies approximately from 2700 kg/m³ for granitic or sedimentary rock to about 3300 kg/m³ for dark mafic rocks like gabbro. Taking an average density of 3000 kg/m³, the vertical stress at the base of a column of sedimentary rock 1000 m thick would be 29.4 MPa. Hence, a sufficiently accurate approximation is that for every 3 km, vertical pressure increases by 100 MPa (100 MPa = 1 kbar = 1000 times atmospheric pressure); thus, ~3 km rock column generates ~100 MPa = 1 kbar vertical stress. The kbar is an old unit but still in common use in geology.

Apart from the arithmetic approximations, there are some complications hidden in the above analysis. Most importantly, the vertical pressure causes elastic distortion that increases lateral pressure above the value expected in a hydrostatic situation. Thus, the lateral stresses (σ_X, σ_Y) cannot be equal to σ_Z as in a truly hydrostatic situation but they relate to it by the ratio of certain elastic parameters of the rock. Whatever prevails and whatever the orientation of the three principal stresses, the mean stress at depth z will be given by the average of the three principal stresses $\frac{1}{3}[\sigma_{MAX} + \sigma_{INT} + \sigma_{MIN}]$ or $\frac{1}{3}[\sigma_X + \sigma_Y + \sigma_Z]$ (Figure 9.1(b)). Any distortion of the rock, whether by fracture or flow, is attributable to the extra tectonic stress known as differential stress $[\sigma_{MAX} - \sigma_{MIN}]$ (e.g., fracture joint; Figure 9.1(c)). This value is usually surprisingly small in the geological environment but acting over geological time intervals, it suffices to move continents. Direct measurements show that typical geological differential stresses are <10 MPa in nature. In room temperature laboratory experiments at a confining pressure of 100 Ma, most rocks require a differential stress of several hundred megapascal before they fracture but elevated temperatures or pore fluid pressures may permit the rocks to flow rather than fracture with differential stresses similar to those measured in nature.

For a given rock, whether fracture or flow occurs depends on primarily on temperature, confining pressure (~depth), differential stress, and also fluid pressure and strain rate, usually in that order of importance. The type of fracture or the type of flow is also dependent on these physical controls. For example, the angle between a shear fracture and the principal stress axes also depends on

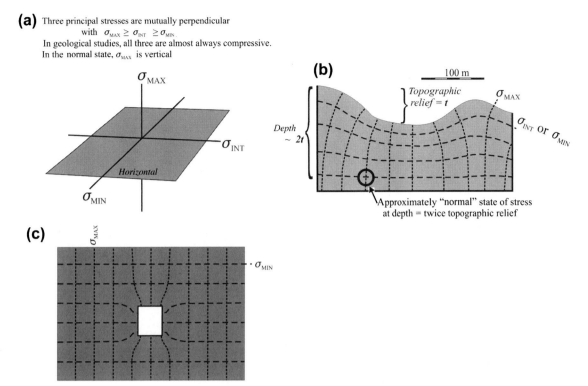

(a) Three principal stresses are mutually perpendicular
with $\sigma_{MAX} \geq \sigma_{INT} \geq \sigma_{MIN}$.
In geological studies, all three are almost always compressive.
In the normal state, σ_{MAX} is vertical

FIGURE 9.2   Stress trajectories.

these parameters, although it is beyond the scope of this text. Of course, the behavior depends on the type of rock also, including its composition, grain size, and anisotropy (=different properties in different directions). Crystalline igneous and plutonic rocks generally resist any type of strain, whereas some sedimentary rocks readily change shape by fracture or flow.

At this stage, reasonable simplified conclusions are that, at geologically interesting depths:

1. Vertical stress (σ_Z) is proportional to the height of the column of overlying rock.
2. If $\sigma_Z = \sigma_{MAX}$, the situation corresponds to Anderson's "normal" state of stress and σ_Z is a function of depth and the density of rock.
3. If $\sigma_Z \neq \sigma_{MAX}$, lateral stresses due to tectonic activity or gravitational sliding from adjacent regions must be operative.

Near the surface, principal stress trajectories must terminate perpendicular or parallel to the ground surface (Figure 9.2(b)). Thus, to a depth dependent on topographic relief, the "normal" state of stress changes and fractures near the surface may curve, following the stress trajectories. Figure 9.2(c) shows how stress trajectories curve around tunnels; fractures may initiate where the trajectories concentrate.

Below the influence of topographic relief or below 2 km, the principal stresses will adopt the normal state, being either vertical or horizontal. Their relationship to igneous

intrusions reveals the simple manner in which tensile fractures relate to the various permutations of principal stress magnitudes and orientations and to the effect of magma fluid pressure (Figure 9.3).

Deeper than 6 km, mean pressure due to the overlying rocks generally suppresses fracturing, essentially "holding the rock together", unless fluid pressures are greater. Fracture mostly occurs in consolidated rock at shallower depths, where rapid movements cause a rate of shape change (generally, strain rate >0.01% per second), or where movement results in more than 1% distortion (=1% strain). Fluids also play a pivotal role in the production of rock fractures; high internal fluid pressures encourage hydraulic tensile fractures, even below 6 km; however, that discussion is beyond our scope.

Since fractures are spontaneous phenomena, usually their orientations correlate with stress axes, which are instantaneous. There are always three orthogonal principal stresses; in geological situations, they are usually compressive. The magnitudes of the maximum, intermediate, and minimum stresses are expressed as $\sigma_{MAX} \geq \sigma_{INT} \geq \sigma_{MIN}$ (Figure 9.4). In Figure 9.4, σ_{MAX} is vertical which corresponds to Anderson's normal state of stress. The planes of maximum shearing stress (τ_{MAX}) bisect the angle between maximum and minimum stress and always contain the intermediate stress axis. Actual shear fractures are not parallel to the planes of maximum shearing stress in consolidated rock rather they occur closer to the σ_{MAX} trajectory

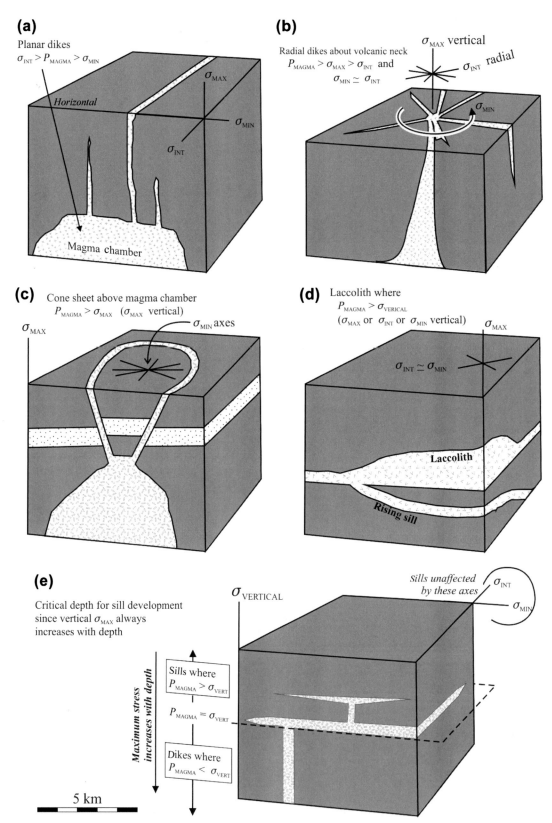

FIGURE 9.3   Tensile failures associated with igneous intrusions.

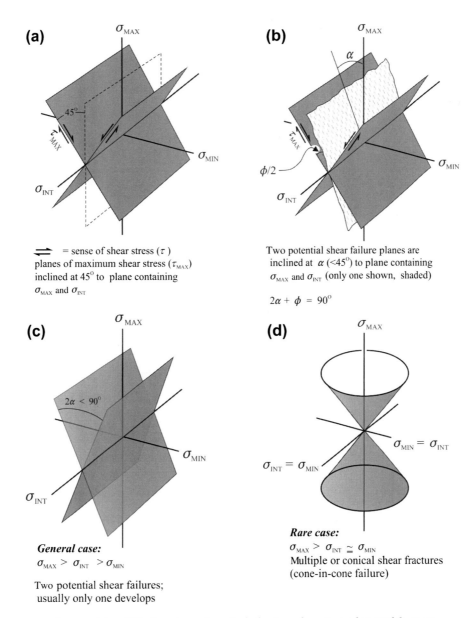

(a)

= sense of shear stress (τ)
planes of maximum shear stress (τ_{MAX})
inclined at 45° to plane containing
σ_{MAX} and σ_{INT}

(b)

Two potential shear failure planes are
inclined at α (<45°) to plane containing
σ_{MAX} and σ_{INT} (only one shown, shaded)

$2\alpha + \phi = 90°$

(c)

General case:
$\sigma_{MAX} > \sigma_{INT} > \sigma_{MIN}$

Two potential shear failures;
usually only one develops

(d)

Rare case:
$\sigma_{MAX} > \sigma_{INT} \cong \sigma_{MIN}$
Multiple or conical shear fractures
(cone-in-cone failure)

*In cases (c) and (d), there is rarely a single fracture plane. Instead several fractures
possess the orientations of the dominant fracture, in different locations in the rock.*

FIGURE 9.4   Shear failure in relation to principal stresses.

(Figure 9.4(b)). The resulting fracture planes usually subtend about 30° with the maximum stress but usually only one possible shear fracture will develop (Figure 9.4(c)). Under rare circumstances, where intermediate and minimum stresses are similar in magnitude, conical fracture systems known as one in cone may develop (Figure 9.4(d)).

When $\sigma_{MIN} > T$, where T is the tensile strength of the rock, a tensile fracture will open, its plane marking the orientation of the σ_{MAX}–σ_{INT} plane (e.g., Figures 9.1(c) and 9.5(a)). A single pulse of tensile failure may typically open a fracture to a width of 2–20 μm. This fracture may mineralize and fill with material of economic importance, precipitated from pore water. Subsequent stress pulses may open

the fracture further; tension joints filled with vein material centimeters wide are very common; this implies many thousands of stress pulses and microfracturing. It is worth noting that veins may occupy large volumes (>25%) of some supracrustal rock sequences, usually where they overly metamorphic rocks that have dewatered. Small-scale shear fractures form shear joints in a similar way (Figure 9.5(b)). They usually occur at a low angle to the maximum stress trajectory.

In learning about stress and fractures, it is useful to recall that the normal stress state (Anderson, 1951; Figure 9.1(b)) is not intended to be universally true nor even precise where its application is valid. It does provide a useful simple

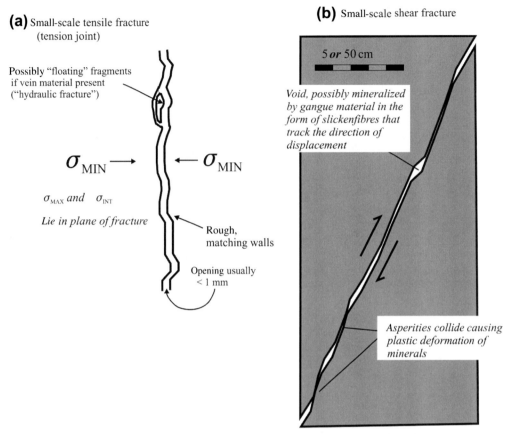

(a) Small-scale tensile fracture (tension joint)

Possibly "floating" fragments if vein material present ("hydraulic fracture")

$\sigma_{MIN} \rightarrow$ $\leftarrow \sigma_{MIN}$

σ_{MAX} *and* σ_{INT}

Lie in plane of fracture

Rough, matching walls

Opening usually < 1 mm

(b) Small-scale shear fracture

5 *or* 50 cm

Void, possibly mineralized by gangue material in the form of slickenfibres that track the direction of displacement

Asperities collide causing plastic deformation of minerals

FIGURE 9.5   Joints (tension and shear) in relation to principal stress trajectories.

explanation for the different orientations of normal, wrench, and thrust faults, with varying permutations of stress axes. The concept goes out from the mathematical basis that any state of stress at a point is defined at any instant, by three principal axes; principal axes are orthogonal and usually differ in magnitude ($\sigma_{MAX} \geq \sigma_{INT} \geq \sigma_{MIN}$). Precisely speaking, there is no such concept as average stress or "Paleogene" stress. The basis for description of the instantaneous stress at a single point in space is the stress tensor; tensors are a much more complicated mathematical concept than vectors; an introduction to stress (and strain) tensors, tailored especially for geologists is given by Means (1976).

Anderson's normal state may be valid near the surface of the Earth, for example, at mining depth in most mining regions. There, the weight of rocks provides the maximum stress σ_{MAX}. Consequently, σ_{INT} and σ_{MIN} must be horizontal; this orientation constitutes Anderson's "normal" state (Figure 9.1(a)). Before we proceed further, let us draw attention to a few key aspects, referring to Figure 9.6.

1. Stress is an instantaneous state; there is no such thing as Cretaceous stress or the stress which caused this fold. A stress state may only be associated with a structure formed by one tiny increment of deformation, usually a fracture.

2. Strain is shape change. Although it is also described by a tensor, it represents the summation of many historical strain increments; it is not an instantaneous feature. We may measure strain now, we may quantify the strain associated with a certain fold or about the strain caused by Cretaceous deformation. The goal of such exercises may be to measure crustal shortening or to determine the strain associated with certain structures such as folds.

3. Stress is almost unrelatable to finite strain (=final total shape change) in very complicated ways that depend on past properties of the rock and the history of orientations and magnitudes of an infinite number of stress states (Figure 9.6(e)). Moreover, the properties of rock that relate stress and strain at any instant are neither homogeneous nor isotropic. (If they were isotropic, the properties would be the same in every direction).

4. Stress axes for any particular past instant are not parallel to the finite strain axes measured now. Neither their magnitudes nor their relative magnitudes are relatable. In geology, it is impossible to define the history of stress increments (axial orientations and magnitudes) that caused a certain finite strain (i.e., produced a certain structure).

Definitive and erudite works encompassing the breadth of structural geology include Ramsay (1967), Ramsay

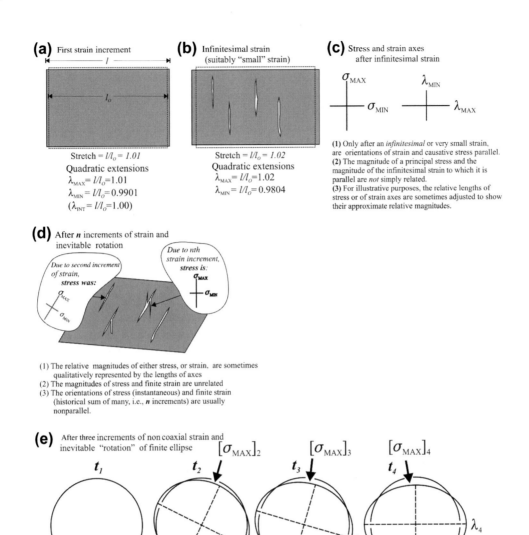

FIGURE 9.6 Simplified explanation of the differences between stress (instantaneous phenomenon) and finite strain (an accumulated history of shape changes).

and Huber (1983), and Ramsay and Lisle (2000). Price and Cosgrove (1990) is both advanced, broader in scope, and yet accessible to the undergraduate. A good introduction to fractures for the geology undergraduate is Price (1966).

Just as mining disturbs the orientations and magnitudes of ground stress, so does a geological fracture. Once a fault or a joint has formed, mean stress $(\sigma_{MAX} + \sigma_{INT} + \sigma_{MIN})/3$ will be reduced and the principal stresses will reorientate. They will be perpendicular to the fracture if it is open but otherwise they will at least be at a higher angle to the fracture after the fracture forms. This relaxed state may be temporary and a renewed level of stress may cause further propagation of the existing fracture rather than fractures. The first fracture usually requires higher stress differences $(\sigma_{MAX} - \sigma_{MIN})$ to form than subsequent fractures.

TENSION FRACTURES

Although most geological fractures are shear fractures, tension fractures are simpler to understand and provide an interesting starting point (Figures 9.3 and 9.5). Tension fractures form where minimum compressive stress fails to hold the rock together, in other words σ_{MIN} > tensile strength of the rock. If a pore fluid exists in the rock, its hydraulic pressure (P_f) works against all the principal stresses, reducing them to effective stresses $(\sigma_{MAX} - P_f, \sigma_{INT} - P_f,$ and $\sigma_{MIN} - P_f)$. P_f may assist the minimum stress so that the condition for failure would be $(\sigma_{MIN} - P_f) > T$. Apart from veins mineralized by fluids, many high-level igneous intrusions invade with the assistance of hydraulic fracturing.

A tensile fracture opens in the direction of the minimum stress axis with little or no motion parallel to the actual fracture. Typically, the fracture is rough, i.e., not very planar, and it may be occupied by vein material or an igneous dike. Thus, the trend of dikes or veins across an area trace out the stress trajectories that operated while fractures propagated. Tension fractures are relatively easily deflected by weaknesses (strength anisotropy) in the country rock. For example, sills may open tensile fractures between beds more easily than dikes that cut across them.

Magma pressure is an effective means of exploiting tensile fracture. Once magma has invaded a fracture, propagation may continue as long as magma flows and as long as its pressure exceeds the tensile strength of the country rock. The simplest example is represented by planar vertical dikes; they crop out perpendicular to σ_{MIN} at the time of their formation and the other two principal stresses lay parallel to the plane of the dike (Figure 9.3(a)). A slightly more complicated situation may exist around a volcanic neck; here, the stress field is not rectilinear. Instead, during volcanic intensity, σ_{MAX} was radial, whereas σ_{MIN} was circumferential. Consequently, radiating dikes may form, a well-known feature in many volcanic areas (Figure 9.3(b)). Another slightly more complex situation occurs where underground magma pressure lifts the country rock to form a ring dike or cone sheet. In this case, the σ_{MIN} stress axes are radial and magma pressure exceeds the tensile strength of the country rock plus its weight, raising a conical block (Figure 9.3(c)). A less dramatic version exists where a body of magma simply bubbles upward to form a laccolith without producing new fractures (Figure 9.3(d)).

A simple further analysis of tensile fracture and stress enables us to identify the conditions for dike versus sill development (Figure 9.3(e)). At depth, dikes form where the magma pressure, P_{MAGMA}, exceeds a horizontal principal minimum stress (σ_{MIN}). At this level $\sigma_{VERT(=MAX)} > P_{MAGMA} > \sigma_{MIN}$. However, at higher levels, the vertical stress decreases to some point at which $P_{MAGMA} > \sigma_{VERT}$; this condition favors sill development, fractures opening horizontally rather than vertically.

SHEAR FRACTURES

Shear fractures are the most common type of brittle failure and they are less easily suppressed by high confining pressure (Figures 9.4 and 9.7). They may occur from the surface to great depths as the textures of ancient faults reveal in rock outcrop. Many shear fractures produce earthquakes but some earthquakes, especially those at very great depth may not be driven by shear failure but rather to phase changes in minerals.

Most brittle failure of rock occurs by shear fractures and it is well understood from laboratory experiment. A shear fracture is inclined to the maximum and minimum principal stress axes and thus has a nonzero shear stress component along its surface, however, the stress component along its surface is not the maximum possible (Figure 9.4(b)). The intermediate stress axis lies in the shear fracture plane. Ideally, the fracture is parallel to the intermediate stress and inclined to the maximum compressive stress at an angle α, where $25° < \alpha < 45°$. This angle does not depend on the state of stress; it is influenced by a rock property known as cohesion or internal friction. Thus, shear fractures do not form parallel to the plane of maximum shear stress (always precisely at $45°$ to σ_{MAX} and parallel to σ_{INT}), which may seem paradoxical at first. Eventually, understanding Mohr's diagram for the stressed state makes this very understandable; the physical property responsible for causing $\alpha < 45°$ is termed the cohesion which is related to a somewhat misleadingly named concept called the internal friction which for most rocks has values in the range $0.3 < \mu < 0.5$. These terms are found in most structural geology textbooks from which a simple generalization is that $\alpha \approx 30°$ (Figure 9.4(b)) or the fault inclines at $30°$ to the maximum stress.

Shear zones are more common at depth, where the confining pressure begins to suppress fracture, usually below 2 or 3 km. Shear zones are a semiductile or ductile equivalent to a shear fracture. Instead of a sharp shear fracture, one finds a zone of sheared and microfractured rock or by a zone of finer grained recrystallized schistose material. Many large wrench faults and thrust faults have a shear zone appearance in their deeper parts.

On maps, most fractures are shear fractures (faults); tension fractures are obviously present where dikes, sills, ring dikes, and cone sheets are present but attention is not normally given to the mechanical aspect. They are simply mapped as igneous sheets. Nevertheless, we should not forget that they have a simply interpreted mechanical origin that may be very useful to us, revealing ancient stress trajectories.

For the field geologist dealing with outcrops and rock specimens, tensile and shear fractures are evident by eye and under the microscope. "Joint" traditionally refers to such small-scale fractures with negligible movement but there is a nontrivial problem of how we distinguish between joints and faults. This really concerns shear fractures and how the amount of displacement and length of the fracture discriminate between joints and faults. Shear joints certainly have small displacements (several millimeters) and small dimensions (lengths of a few meters), whereas faults have displacements of measured at least in meters and lengths at least in tens of meters. The distinction is somewhat vague and arbitrary in many instances and informally, many geologists would carry in their minds upper and lower limits for the reasonable lengths of shear joints as opposed to faults. Surprisingly, information on the dimensions of fractures and their displacements are difficult to obtain and correspondingly scarce. One commonly quoted data set implies that for most map scale shear fractures faults, displacements on average may represent 3% of the total fault length. Such estimates are broad generalizations subject to the type of fault, conditions of faulting, crustal level, and how the "length" is defined with respect to the slip direction along which displacement is recorded.

The underlying cause of faults in decreasing order of scale is usually, plate tectonic motion, uplift of the crust due to denudation, crustal loading due to sedimentary deposition, and igneous activity. Earthquakes are one manifestation of the initiation and incremental movement of faults or joints. What most geologists describe as joints invariably owe their origin to minor local tectonic movements, regional stress fields, and crustal uplift. They are confined to the upper few kilometers of the continental crust and are present to shallower depths within the oceanic crust. Joint orientations, frequency, and motions generally relate to local features (folds, nearby faults) or to subregional stress fields.

SIMPLIFIED MECHANICAL DETAILS

One practical consequence of a structure that forms by brittle failure is that it will not propagate across a free surface (Figure 9.6). Thus, if fracture Y_F is younger, it may stop at an older fracture O_F because the required stresses cannot be projected across the tiny gap caused by O_F. From a map, it may thus falsely appear that O_F truncates Y_F. Thus, with intersecting fractures some exceptions to the law of "cross-cutting relationships" may occur.

Joints are rarely large enough to present on a geological map, although their orientation may be recorded at individual locations by special symbols. The most important differences between joints and faults are that joints involve negligible displacements, often less than a millimeter. Also joints may form either as a result of tensile failure, with the fracture perpendicular to minimum compressive stress (σ_{MIN}) and parallel to (σ_{MAX}), or due shear failure where the fracture plane is inclined to σ_{MIN} and σ_{MAX}. Note that any state of stress is ephemeral and always described by three orthogonal principals stresses: $\sigma_{MAX} > \sigma_{INT} > \sigma_{MIN}$. Where $\sigma_{MAX} = \sigma_{INT} = \sigma_{MIN}$, no deformation is possible and the state of stress is described as hydrostatic. Lithostatic is the geologist's synonym for hydrostatic pressure caused by the burial of rock; however, it is an imperfect analogy since rocks are not Newtonian fluids, even when considered on geological time scales, and thus always bear significant shear stress (differential stress).

In contrast, faults have substantial displacements from meters to hundreds of kilometers and they always result from shear failure, commonly with the fracture surface inclined at about 30° to σ_{MAX}. It is interesting and important to note that fractures are among few geological structures that may be directly and simply related to the orientations of stress. Almost all other secondary structures result from the

accumulation of many increments of shape change, called strain increments. They may reveal the orientation of finite strain axes that are quite different from stress axes.

An individual increment may cause 0.5% extension of the rock and the maximum extension would be parallel to the minimum compressive stress, σ_{MIN}. If a shear fracture formed during this increment of strain, its orientation could be related directly to the orientation of the principal compressive stresses; $\sigma_{MAX} > \sigma_{INT} > \sigma_{MIN}$. All things being equal, two possible orientations are possible for the shear fracture, occasionally, both may form more-or-less synchronously. They are the conjugate orientations. The shear fracture always contains the σ_{INT} axis and it is commonly inclined at about 30° to σ_{MAX}. (This orientation is not the maximum shear stress plane as one learns in structural geology.)

The practical value of the stress trajectory to fault relationships is that they lead to a rather simple and usually correct classification of faults, attributed to Anderson (1951). Anderson recognized that in most of the continental crust the normal state of stress was that σ_{MAX} was nearly vertical due to the weight of rock. A simple hydrostatics analogy shows that the vertical stress increases with the depth and is proportional to the density (ϱ):

$$\sigma_{VERT} = \rho g z,$$

where g is the acceleration due to gravity and z is depth. This implies that average crust ($\varrho = 2750\,kg/m^3$) causes a vertical pressure of approximately 1 kbar (100 MPa) per 3 km. We now know that the normal state of stress:

$$\sigma_{MAX} = vertical;\ \sigma_{INT} = horizontal;\ \sigma_{MIN} = horizontal,$$

also prevails over most of the ocean crust, excluding the vicinity of island arcs and oceanic trenches (subduction zones). For oceanic rocks, we must take account of the weight of the water column also which depends on the depth of the ocean floor at the time of faulting. Since most oceanic faulting occurs early in ocean floor spreading, near the mid-ocean ridge, depths may be relatively shallow compared with the ~4 km deep abyssal plains. Faulting near the deep oceanic trenches occurs under deep water columns, of course. For seawater ($\varrho \approx 1030\,kg/m^3$), hydrostatic pressure is $\approx 10\,MPa$ ($\approx 100\,bars$) at a depth of 1 km.

Anderson noted that the most common faults in the continental crust dip toward the direction of the dropped or downthrown side. Associated with the normal state of stress (σ_{MAX} vertical), they are designated normal faults. Near mountain belts and major shear zones where tectonic stresses inclined gently to the earth surfaces dominate thrust faults abound. These are associated with σ_{MIN} vertical and σ_{MAX} horizontal. Since the fracture surface is inclined at ~30° to σ_{MAX}, it is evident that most normal faults dip at ~60°, whereas most thrust faults dip at ~30° (see diagrams).

(a)

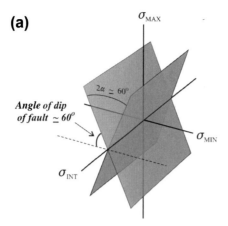

Anderson's "normal state of stress", common in the shallow crust has maximum compressive stress nearly vertical, and this is attributed to the weight of the rocks.

The fault most likely to develop in this system is therefore termed the "normal fault". As a consequence of the stress orientations, it dips at ~60° [= 90 − α]. One of the two equally probable fault orientations will develop.

(b)

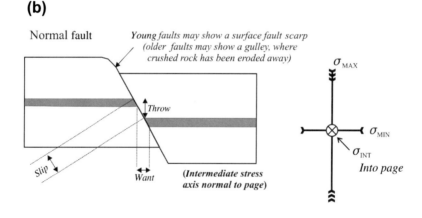

FIGURE 9.7 Definition of normal fault.

Occasionally, faults dipping at ~60° do not dip toward the downthrown block. Since the fracture must have formed in a normal state of stress (Anderson), this is described as a reverse fault; a fault initiated as a normal fault that has reversed its motion due to later subhorizontal compression.

Where σ_{INT} is vertical, the fracture surface must also be vertical. These are named wrench faults (synonyms are tear, lateral, and transcurrent) and their displacement is viewed directly in the plan view of a map. Since the displacement is sideways, there is no throw (vertical motion). The fault is a right or left wrench fault, depending on whether the opposite fault block moves to the right or left with respect to the fault block in question. The Latin terms dextral (right-handed) and sinistral (left-handed) are very commonly used. Wrench faults are readily identified on maps since they usually preserve their original vertical orientation. Wrench faults have the largest displacements of any

continental fault, sometimes hundreds of kilometers accumulated by millions of small movements.

The simplest and the most common fault is the normal fault, resulting from the "normal" state of stress (Figure 9.7). The fault dips at about 60° due to the previously described mechanical principles. The normal fault dips toward the downthrown side and leaves an area off "want" underground where horizons are discontinuous (Figure 9.7(b)).

The other classes of fault: thrust (Figure 9.8(a)), reverse (Figure 9.8(b and c)), and wrench (Figure 9.8(d)) are attributable to the three possible orientations of principal stresses.

EXPRESSION OF FAULTS AT THE SURFACE

The surface offset produced by faults with a vertical displacement seems paradoxical at first seem until one considers the combined effects of displacement followed by erosion. Usually, the faulted topography becomes eroded

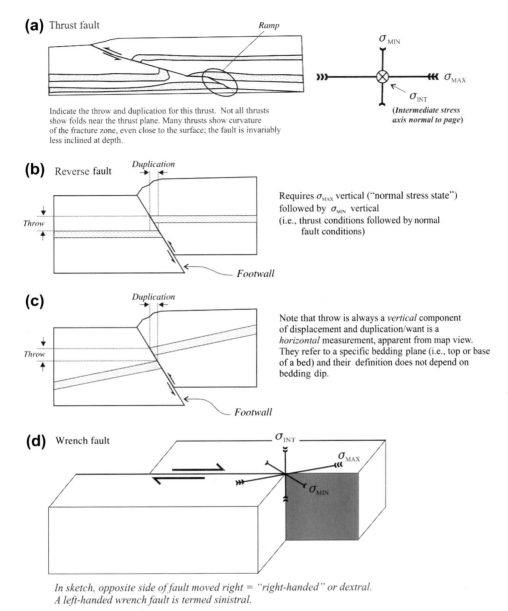

(a) Thrust fault

Ramp

σ_{MIN}

σ_{MAX}

σ_{INT}

(Intermediate stress axis normal to page)

Indicate the throw and duplication for this thrust. Not all thrusts show folds near the thrust plane. Many thrusts show curvature of the fracture zone, even close to the surface; the fault is invariably less inclined at depth.

(b) Reverse fault

Duplication

Throw

Footwall

Requires σ_{MAX} vertical ("normal stress state") followed by σ_{MIN} vertical (i.e., thrust conditions followed by normal fault conditions)

(c)

Duplication

Throw

Footwall

Note that throw is always a *vertical* component of displacement and duplication/want is a *horizontal* measurement, apparent from map view. They refer to a specific bedding plane (i.e., top or base of a bed) and their definition does not depend on bedding dip.

(d) Wrench fault

σ_{INT}

σ_{MAX}

σ_{MIN}

In sketch, opposite side of fault moved right = "right-handed" or dextral.
A left-handed wrench fault is termed sinistral.

FIGURE 9.8   Definition of thrust, reverse, and wrench faults.

(Figure 9.9) so as to appear to "displace" the outcrop of dipping beds laterally. Figure 9.9 should be studied carefully to appreciate these points and the complications of a dipping fault should be noted also (next section, Figure 9.10).

DIP-SLIP AND STRIKE-SLIP COMPONENTS OF MOTION ON A FAULT

In the following exercises, for simplicity, faults will have either vertical or lateral motion (Figure 9.10). In reality, faults with components of vertical and lateral motion are common but in detail, some combination of strike-slip and dip-slip motion may occur (Figure 9.10). However, if the vertical motion is negligible in comparison with the lateral component, the fault would simply be mapped as a wrench fault. An accompanying map report may note in detail that some vertical motion was present, based on some field or evidence.

In the general case, both vertical and horizontal displacements could be significant although one will usually dominate. The motion on the fault plane is described by a slip vector; a direction on the fault plane with a magnitude describing the displacement in that direction (Figure 9.10). It is an important parameter but it is usually determined from the slip components that may be more easily determined from a map or in the field. These are the strike-slip (ss) component of motion, the displacement in a horizontal

(a) Vertical motion causes *apparent* lateral offset where

(1) Vertical or dipping fault.
(2) Only requires vertical displacement (throw); no lateral motion on fault.
(3) At any level, e.g., mine plan or a later erosional surface, any nonvertical feature will be offset laterally across the fault This is not a real lateral displacement.
(4) The throw may be determined only by comparing structure contours across fault.

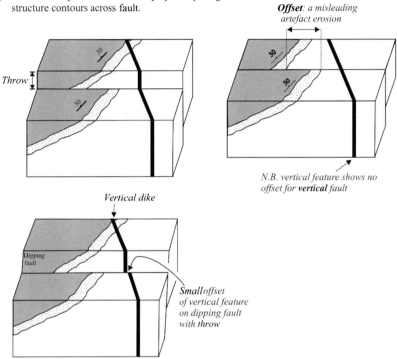

(b) True lateral motion

(1) The fault need not be vertical.
(2) There may be both true lateral motion and throw.
(3) If there is *only* lateral motion, offsets in map view do reveal displacement accurately.

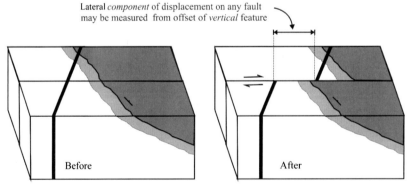

FIGURE 9.9   The topographic expression of faults.

line on the fault plane, and the dip-slip (ds) component, the displacement along the dip direction of the fault plane. Referring to the diagram, it is evident that where the fault dip angle $= \theta$ and where the slip vector makes an angle α with the strike line (in the plane of the fault) and where the throw is t:

so that

$$ds = \frac{t}{\sin \theta} \text{ and } ss = \frac{ds}{\tan \alpha},$$

$$slip = \sqrt{(ds^2 + ss^2)}.$$

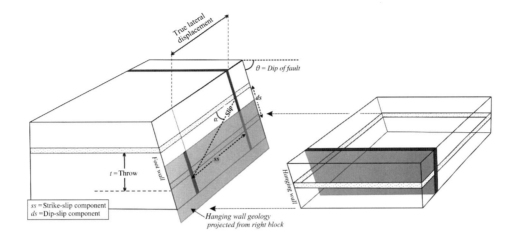

The fault shows both dip-slip and strike-slip components, which may have occurred simultaneously.

In nature, almost every fault has a component of slip along the dip direction of the fault plane (*dip-slip*) and some component of lateral slip (*strike-slip*). On the map scale, one of these components is usually negligible. Nevertheless, the general and the most complex case is shown here:

A dipping fault with significant dip-slip and strike-slip components.

For visual simplicity, the relationships are illustrated using a vertical dike and horizontal bed. Without a vertical feature, is it not possible to measure the true lateral displacement without using structure contour constructions. Map view "offsets" are even more misleading and complex for faults with dip-slip and strike-slip displacements.

FIGURE 9.10 Dip-slip and strike-slip components of fault displacement.

We know *t* and *θ* from map constructions; it may be possible to determine ss directly from the map rather than by the intermediate calculation shown above. Thus, from a map alone, it may be possible to determine the true amount of slip on a fault. Note that a vertical feature will show apparent lateral displacement on a dipping fault.

TRANSFORM FAULTS AND OTHER GROWTH FAULTS

Although our map problems will not deal with transform faults, we should consider them, if only to avoid any confusion with wrench faults that appear beguilingly similar (Figure 9.11). Transform faults are a special class of fault originating at mid-ocean ridges where ocean floor creates basaltic magma in vertical dikes by that strike parallel to the ocean ridge (Figure 9.12). Along the spreading axis, the magma feeders are not equally active all the time, causing the ocean floor to spread at different rates along the ridge. To accommodate this differential growth of new ocean crust, transform faults, first recognized and understood by Professor T. Wilson (University of Toronto), separate offset segments of the expanding spreading axes, which may spread at different rates.

This is a special kind of growth fault, in which ocean ridge offsets are in the opposite sense to the motion but only occur on the active inter ridge portion of the transform fault (Figure 9.12). Motion is restricted to the part of the

transform fault between the ridge where ocean floor spreading directions are opposed. For example, some of the earliest recognized examples, south of Baja California, show apparent sinistral offsets of the spreading axis but the motion on the active portion of the transform fault is actually dextral. Where transform faults pass under the continental margin, they affect the continental crust as wrench faults and the underlying transform nature is not apparent from continental geological data alone. However, as a rule, most long wrench faults are extensions of transform faults from an oceanic terrane. For example, the dextral san Andreas system is an extension of the offshore dextral transforms south of Baja California. Other onshore wrench fault expressions of transform faulting include the New Zealand Alpine Fault, the Japanese Median Tectonic Line, and part of Holy Lands (Dead Sea) Fault system. In more ancient terranes, there may be no surviving evidence of the ocean floor but for wrench faults with huge displacements, there is really no sensible alternative to a transform origin. For example, a transform origin is mandatory for the Great Glen Fault of Scotland, which is traceable from Shetland through Scotland to Northern Ireland and which correlates with the Fleur de Lys Fault of Newfoundland; a reconstruction dependent on pre-Atlantic opening.

Another type of growth fault provides a similar problem, although less commonly. Some large normal faults, especially flanking rift valleys (=Graben), accumulate displacements progressively as sedimentary or volcanic deposition

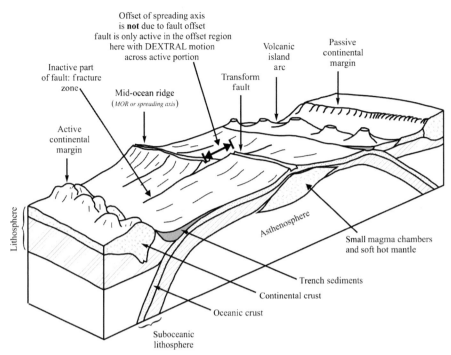

FIGURE 9.11 Transform faults are a special class of growth fault, mainly found under the oceans. Only a part of their length, near the mid-ocean ridge, is active.

occurs on the downthrown side. In fact, deposition may sustain the faulting process and it may be different in character on either side of the fault since the depositional environment may differ. It may be difficult to establish a unique value for the throw since strata of equivalent age will be thin on the hanging wall side and thicker on the footwall and possibly differ in appearance. Commonly sediments on the footwall side may be coarser and even involve slump breccias due to the steep depositional slopes caused by syndepositional faulting. Classic examples of growth faults are provided by most major rift valleys: Cenozoic examples are well known from Kenya, the Dead Sea, and Polis Graben Cyprus; Paleozoic example of Midland Valley of Scotland; and Proterozoic example of the Nipigon embayment in northern Ontario. Large normal growth faults are sometimes termed half-grabens in the literature but this seems unnecessary jargon. They are simply large normal faults that control large differences in strata thicknesses and perhaps sedimentary character; examples are the Polis Fault (Cyprus) and Church Stretton Fault (UK) (Figure 9.20).

A more complex and irregular growth fault situation may arise along the active portion of some transform faults (Fig.9.11). As the wrench displacement occurs, parts of the seafloor with different elevation are inevitably juxtaposed, giving an effective dip-slip component of displacement. These slopes generate submarine breccias and provide local basins for exotic off-ridge volcanogenic sediments that accumulate on the low side of the fault. Which side of the fault is low depends on the submarine topography. This

is difficult to study unless ocean floor rocks crop out land as ophiolite; an example in Cyprus is the Cretaceous Arakapas fault, part of the southern Troodos Transform Fault. The presence of transform-fault-margin sedimentation is evident under the modern oceans since those areas provide sources of readily dredged fragmental basalt. Active slip occurs only between the ocean ridge segments and the inactive extensions of the fault are termed fracture zones. It is always important to remember that the actual active faulting motion is due to differential growth of the MOR and in the example shown, it is dextral motion. Most long wrench faults on land are extensions of offshore transform faults, for example, the San Andreas Fault results from a transform passing under the coast of California.

REVIEW QUESTIONS CONCERNING FAULTING

1. What are the most common dip angles for normal, reverse, thrust, and wrench faults?
2. Which type of fault is the most common in the shallow crust?
3. Where are transform faults found and how do they differ from wrench faults?
4. Which type of fault is most common along the margins of active mountain belts?
5. Which type of fault may have the largest displacement?
6. Which type of fault may have the second largest displacement?

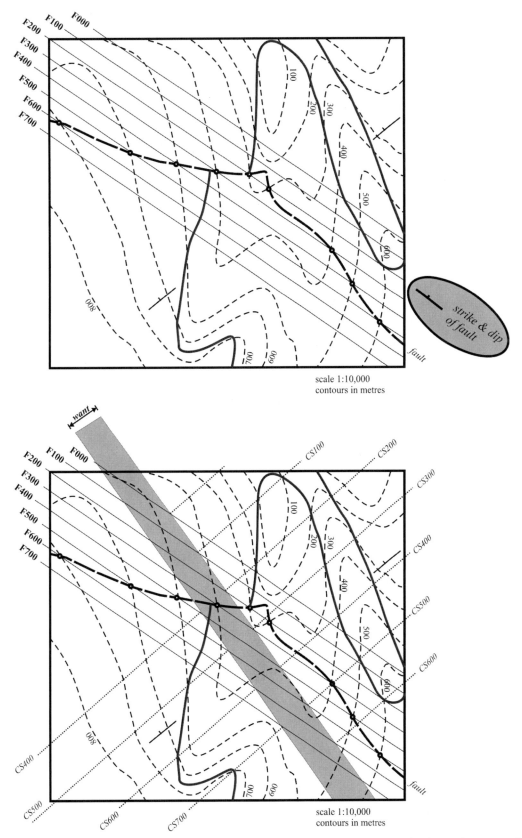

FIGURE 9.12 Structure contour procedure to determine displacement on fault. (a) partial construction (b) complete construction.

7. How do the angles of dip of normal and thrust faults usually change with depth?

SIMPLIFIED PROCEDURE FOR UNDERSTANDING A FAULT FROM A MAP

The approach to understanding faults from mapped information may safely and most effectively proceed in the following manner (Figure 9.12).

1. What is the dip of the fault? If the topography is perfectly flat, any planar fault will crop out as a straight line and there is no topographic clue to its dip. This is unlikely, however, the curvature taken by the fault outcrop will indicate the dip; the more sinuous its, the shallower is the dip of the fault. Fault dip, if unmodified by subsequent earth movements, is a clue to the type of fault but it is not foolproof. Generally, normal and reverse faults dip at about 60°, thrust faults dip at about 30° and wrench faults are mostly vertical. The angle is determined by constructing structure contours for the fault plane but with practice, one may gain a fair estimate of the dip direction by visual inspection of the fault's outcrop and the pattern of topographic contours.

2. What is the sense and amount of displacement? This is the definitive manner in which to classify the fault. It may only be reliably determined by constructing structure contours for the same horizon on each side of the fault. If the sense of motion is vertical, downthrow in the direction of fault dip defines a normal fault. Upthrow of the fault block away from the direction of fault dip defines a thrust or reverse fault. Hanging wall and footwall are also be used to refer to the parts of the upthrown and downthrown blocks adjacent to the fault plane. This knowledge can only be obtained from constructing structure contours for the strata and terminating them on the structure contours for the fault (e.g., Figure 9.12).

3. Appearance of displacement from the map view is usually very misleading, to the disappointment of novices. As illustrated in the diagrams, postfaulting erosion and topographic effects conspire to give quite false impressions. The exception occurs where a vertical feature (e.g., dike) is faulted by a vertical fault; in this special case, any lateral displacement is faithfully recorded in map view. Note that where a vertical feature (e.g., dike) is obliquely faulted by a dipping fault it generally shows a false lateral motion (Figure 9.10).

Figure 9.12 shows the steps used in studying a fault on a map. In the upper copy of the map, has structure contours for the fault plane. The contours are equally spaced so we know that the fault is planar and has uniform dip. The spacing of the structure contours enables us to determine the fault dip; this is shown on the right hand side of the map.

In the lower copy of the map, structure contours are shown for a single bedding plane, in this case a coal seam (CS). The contours are constructed for each side of the fault separately. The structure contours are terminated on the fault where a contour for the fault has the same value. For example, CS200 terminates on the contour for the fault F200. We then proceed to construct structure contours on the SW side of the fault but again we terminate them against the fault contour off the same value. For example, CS400 terminates on F400.

We now see that CS200 on the NE side of the fault lines up with CS400 on the SW side of the fault, indicating that the throw of the fault is 200 m down to the NE. Since the throw is down to the NE and the fault dip down to the NE, we are dealing with a normal fault. This is the only reliable means of determining the throw of a fault. Inspection of the map and outcrop patterns alone does not provide a definitive answer. Note the area of want; beneath which the coal seam is missing due to faulting. All normal faults produce a band, beneath which horizons are absent (see Figure 9.12(b)). Conversely, reverse and thrust faults produce an area of duplication, beneath which strata duplicate (see following exercises). Complete the map (Figure 9.13) by drawing a cross-section from the SW corner to the NE corner and indicate the dip angles (on the map) for the fault and for the beds.

Normal Fault

Treat the map as with Figure 9.12. The tick indicates the downthrown side of the fault. First draw structure contours for the fault (Figure 9.13). Then, draw structure contours for the bedding plane on each side of the fault, separately. Terminate the structure contours for the bedding plane on the fault and define the area of want. Draw a cross-section using the section given. Note the absence of the dike in the section. Explain this. Determine the throw of the fault from structure contours. You cannot determine the throw from the cross-section.

Reverse Fault

Treat the map as Figure 9.13 but note the fault dips steeply toward the upthrown side. Determine the throw of the fault, its area of duplication and explain the effect on the dike (Figure 9.14). Can a reverse fault be formed during a single episode of fault movement?

Draw the cross-section.

Wrench Fault

Since the wrench fault is vertical, the offset of the vertical dike gives the displacement (d) accurately (Figure 9.15). The sense of displacement is dextral (right-handed) as shown by the arrows. The opposite sense of displacement is sinistral (left-handed). In the absence of the vertical dike, the throw

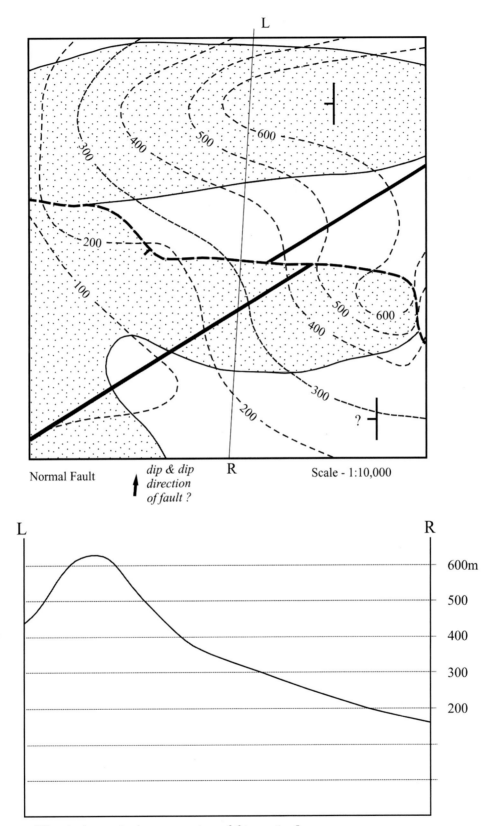

Normal Fault   dip & dip direction of fault ?   Scale - 1:10,000

What is the vertical exaggeration of this section?

FIGURE 9.13 Normal fault.

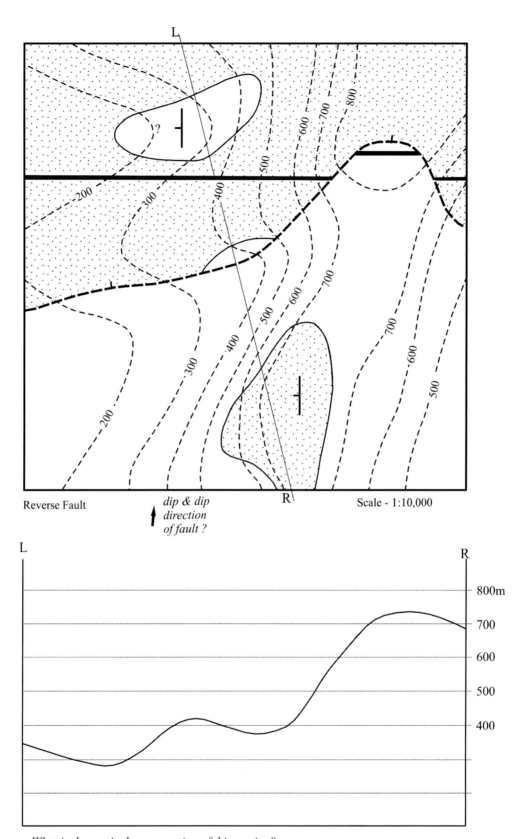

Reverse Fault

What is the vertical exaggeration of this section?

FIGURE 9.14   Reverse fault.

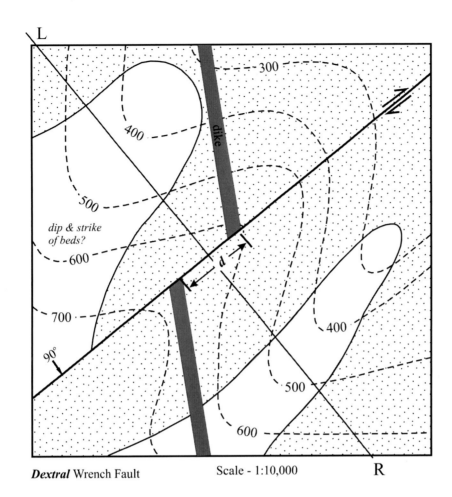

Dextral Wrench Fault   Scale - 1:10,000

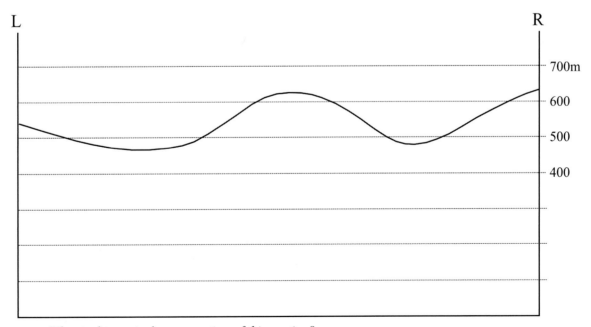

What is the vertical exaggeration of this section?

FIGURE 9.15 Wrench fault.

would be determined by constructing structure contours on either side of the fault and comparing them where they meet on the fault. Do this and compare the answers for the throw. Draw a cross-section and explain the absence of the dike. (This is equivalent to the want of a normal fault.)

Thrust Fault

Since thrust faults generally dip gently (30° or less), they follow tortuous paths across the map, affected considerably by topography (Figure 9.16). The barbs on the thrust point to the overlying block. Determine the dip of the beds and of the fault. Terminate structure contours for the bedding plane on the fault at the appropriate elevations. Determine the throw of the thrust and draw the cross-section. Note the large area of duplication.

Fault Problem

Draw structure contours for the fault and determine it dip. Choose one bedding plane and construct structure contours, on each side of the fault, separately (Figure 9.17). Determine the throw of the fault. What type of fault is this? Draw cross-sections LR (left to right) along lines 1 and 2. Explain and illustrate the effects of faulting on the dike.

Curvature of Faults

Many faults show a curvature, due to the curvature of the causative stress trajectories. This is noticeable with thrusts (a) and to some extent with normal faults (b). Figure 9.18(c) maps a sandstone unit in fault contact with shale. Choose one sandstone bedding plane and construct structure contours. They are parallel but not equally spaced. What does this tell you about the fault? What kind of fault is this?

Runcorn Coalfield, UK

In coalfields, mapping is commonly very detailed and particular attention is paid to the location and displacement of faults (Figure 9.19). This example illustrates the runcorn coalfield of northern England. Dips and strikes of strata are shown and the downthrown side of the faults is shown with a tick; in some cases, the amount of throw is given in meters. A graben is a down-faulted portion, between two normal faults. Identify several graben (rifts). Try to determine the relative ages of the named faults. Finally, draw a cross-section from L (on the left) to R, ignoring topographic relief, which is very subdued. (Return to this problem later to plot a rose diagram of the fault strikes, see Figure 9.24).

Midwales

Determine the dip direction and strike direction of the strata at points indicated with a solid dot (Figure 9.20). (It is not possible to determine the dip angles.) In view of the scale of the map, the topographic relief is negligible. Indicate the downthrown side of the faults marked "?" from stratigraphic considerations. Do the Pre-Cambrian metamorphic rocks form a graben or a horst?

Mark the positions of significant angular unconformities on the map and in the stratigraphic column. Draw a cross-section AB with A on the left. Tabulate the geological history.

Czech Republic–N. Slovakia

Again, topographic relief is negligible because of the scale but some prominent mountains are shown for location (Figure 9.21). Draw a cross-section N–S with north on the left showing the main faults and their sense of displacement. Indicate the orientation of beds or schistosity where you have information bearing in mind that schistosity is commonly close in orientation to thrusting. Finally, on the map, indicate the locations of angular unconformities by tracing them out. Thrust sheets are also termed thrust nappes from the French "nappe" for a sheet or cloth. In the next chapter, we shall eventually learn that the strata in thrust sheets may also be tightly recumbently folded to give fold nappes.

Poland–Slovakia

Topographic relief is negligible in comparison to the scale (Figure 9.22). Trace out the basal unconformity of posttectonic sedimentary rocks (mollasse). Draw a cross-section AB with A on the left including approximate orientations of thrusts and schistosity.

Dead Sea Rift (Transform)

Faults are not shown in this map (Figure 9.23). You may discover them in structural geology textbooks or from the web or infer them here from the stratigraphy. Mark the east and west boundary faults to the rift, to the best of your ability, and draw an EW cross-section. The east and west boundaries to the rift overlie an N–S striking transform fault.

Use of Rose Diagram to Show Fault Orientations

The distribution of strikes of faults may yield useful information on stress history or simply on the relative ages of faults (Figure 9.24). Figure 9.24(a) shows a histogram of fault and joint orientations but this is not as helpful as it could be since the attribute of orientation is not clear. Wrapping the histogram around a circle (Figure 9.24(b)) reveals which trends are most popular in a meaningful way. Figure 9.24 (d and e) provides blank rose diagrams which you may use for your own projects, such as the orientation distribution of faults in the runcorn coalfield (Figure 9.19).

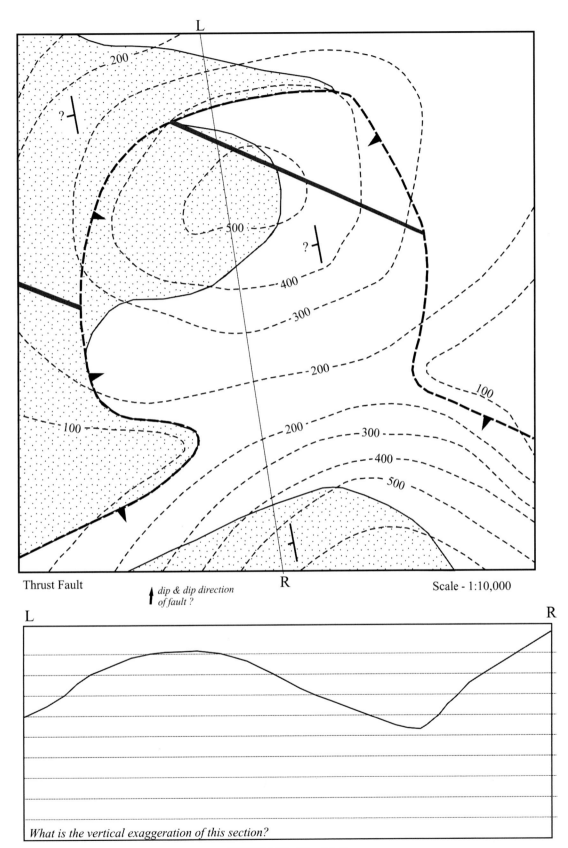

Thrust Fault

↑ *dip & dip direction of fault ?*

Scale - 1:10,000

What is the vertical exaggeration of this section?

FIGURE 9.16 Thrust fault.

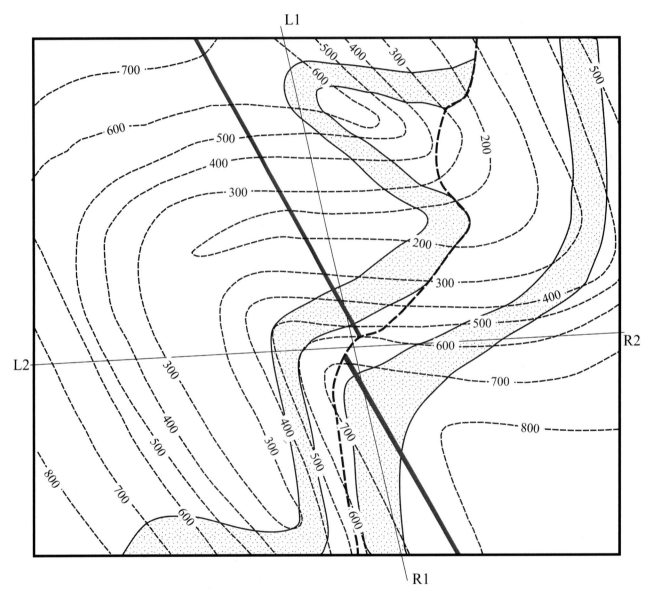

Scale 1:10,000 Elevations in metres

FIGURE 9.17 Reverse fault.

(a) Change in thrust-fault dip due sympathetic with maximum stress trajectories

(b) dip-change of Normal-fault due to increase in stress difference [$\sigma_{MAX} - \sigma_{MIN}$], with depth

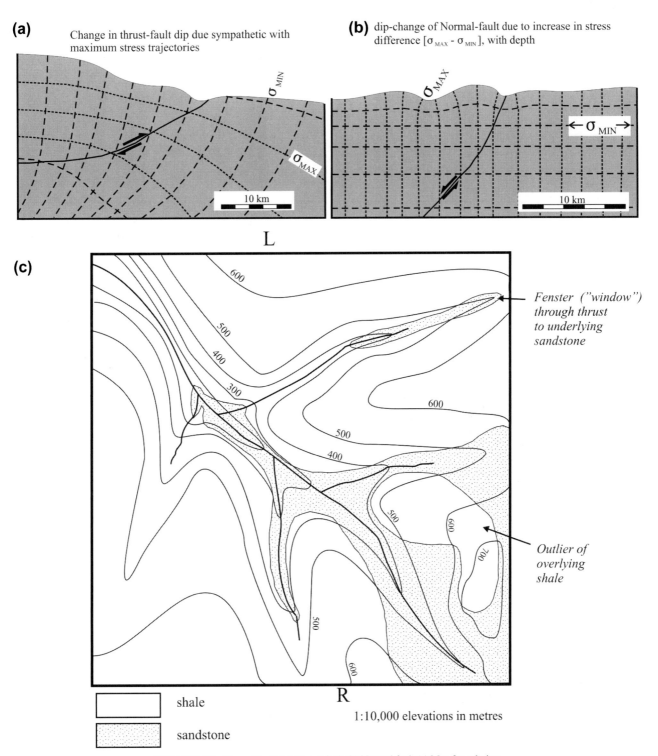

(c)

L

R

Fenster ("window") through thrust to underlying sandstone

Outlier of overlying shale

shale

sandstone

1:10,000 elevations in metres

FIGURE 9.18 Curved fault. (a) Thrust fault (b) Normal fault (c) Map for solution.

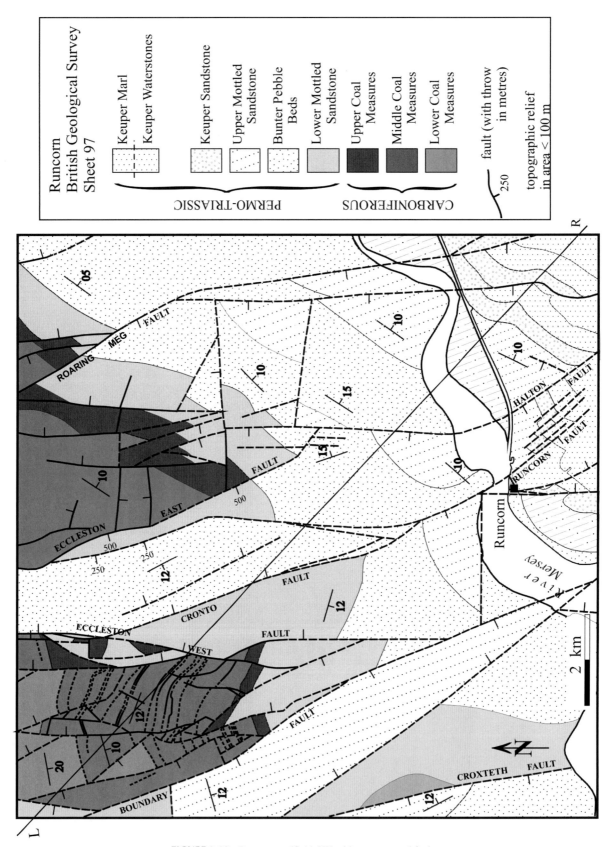

FIGURE 9.19 Runcorn coalfield, UK with many normal faults.

Mid-Wales & Marches
Sheet 52°N - 04°W

FIGURE 9.20   Midwales, UK; faults and folds.

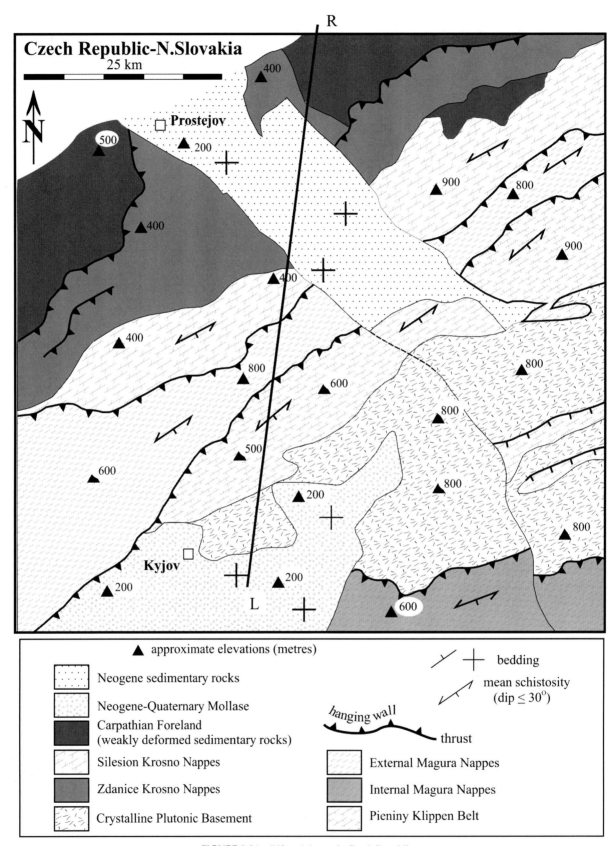

FIGURE 9.21 Rift and thrusts in Czech Republic.

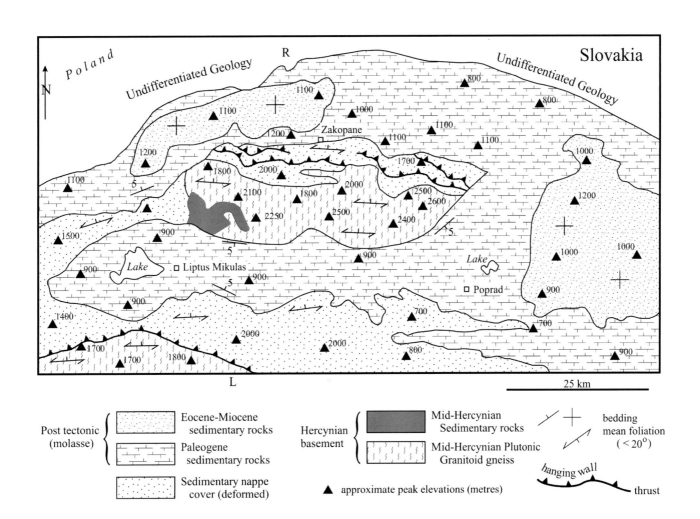

FIGURE 9.22 Thrusts in Slovakia.

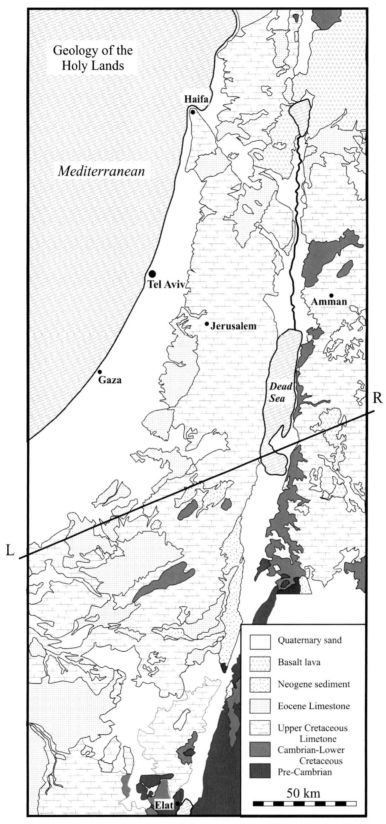

FIGURE 9.23 Dead Sea rift and transform, Israel. Two NS transform faults bound the Dead Sea. Their differing dip-slip components cause a rift to form. Inferring from stratigraphy and by research in structural geology textbooks locate the faults and the rift. Which side of the rift is most downthrown?.

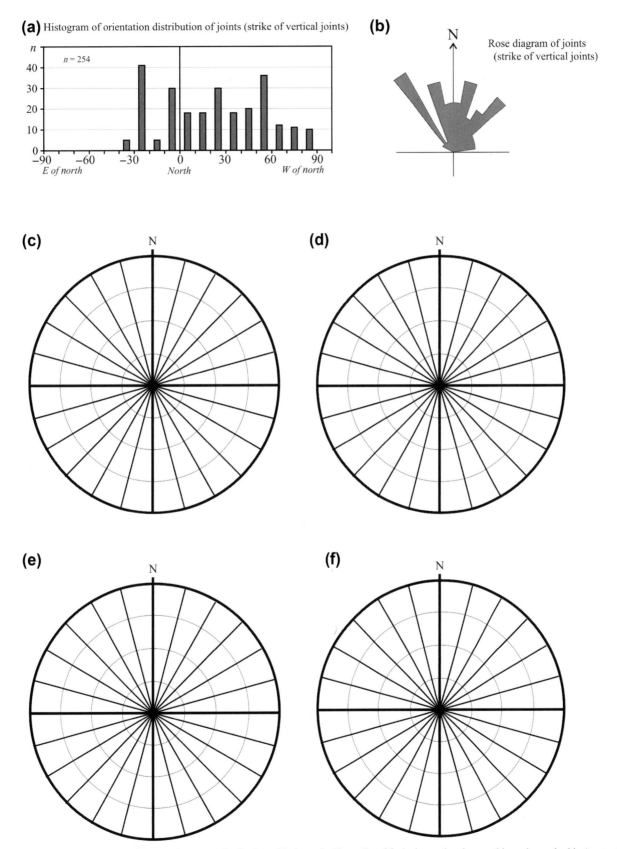

(a) Histogram of orientation distribution of joints (strike of vertical joints)

(b) Rose diagram of joints (strike of vertical joints)

(c)

(d)

(e)

(f)

FIGURE 9.24 Rose diagram to illustrate frequency distribution of fault trends. The strike of faults is noted and counted in each angular bin (see text). (a) Histogram of joint frequency (b) Rose diagram of same data (c-f) blank rose diagrams for use (e.g. Runcorn map).

Folds

When a geology student confronts a geological map for the first time, the geological boundaries may appear as a bewildering pattern. Soon we learn to recognize that the contortions and bends of boundaries are due the topographic surgery of the solid geology, especially if the map is at a small scale. Even a horizontally bedded sequence shows a complex outcrop pattern; it follows the contortions of the topographic contours precisely. Why is it that we are so comfortable with topographic maps and topographic contours yet initially perplexed when confronted with a geological map? Like the pioneer, William Smith, we learned to interpret the interference of geological layers with the topography in order to understand their relative ages. Perhaps by now we realize Smith's genius. To paraphrase Charles Darwin, we do not understand simply by looking or sketching; we understand truly when observations are perceived by the "inner eye of reason". We often take important simple ideas for granted; but remember that before someone introduced them there was usually ignorance or misunderstanding.

So far, we discussed strata whose map patterns are sinuous due to topography. We understood in those examples that the strata were essentially planar and not actually contorted. Now we meet cases where the strata are contorted; they exhibit folds, undulations of the layers due to tectonic movements that usually commence with the process called buckling by structural geologist. Other processes, usually shortening, often subsequently amplify the buckles. Buckling has the attractive property that the undulations of layers have an approximately regular waveform as long as the layers have constant thickness and uniform lithology. Thus, geologist will talk of the wavelength and amplitude of folds, as though they were sinusoidal waveforms in physics. Of course, such precision is lacking in nature

and some folding processes produce highly irregular nonperiodic undulations of layers, even if they are of uniform thickness and lithology (e.g., disharmonic folds, ptygmatic folds, etc. in high-grade rocks). Since any folds produce different changes in orientation in different parts of the bedrock they are considered as examples of heterogeneous strain. Homogeneous strain occurs where straight lines remain straight and parallel lines remain parallel. Homogeneous strain is rather rare and usually only approximated at some certain scale. Every rock is heterogeneously strained at some scale or other.

Folds may occur in soft sediment due to slumping. More commonly, we recognize them as the result of tectonic movements in sedimentary rock, in metamorphic rock of every grade, and in some layered igneous rocks. Folding is a common response of rocks and sediments to motions slower than those involved with faulting and usually with stresses that are much lower than those causing faulting. Generally, they result from steady-state metamorphic flow that also produces schistosity or cleavage approximately parallel to the axial planes. A fabric lineation may also be produced parallel to the maximum extension direction that is normally at a high angle to fold axes. A fold is a wrinkle whose alignment or axis is perpendicular, or nearly so, to the shortening direction. Axes are not necessarily perpendicular to the maximum compressive stress since folds evolve slowly; in contrast, a state of stress is an instantaneous phenomenon. Folds accumulate increments of strain over a long period to produce a state of heterogeneous finite strain. Thus, as explained in Chapter 9, finite strain structures such as folds cannot be related to stress axes, which are instantaneous changing phenomenon (Figure 9.6).

Understanding Geology Through Maps. http://dx.doi.org/10.1016/B978-0-12-800866-9.00010-7

Calculations suggest that cliff-sized folds along the Devonshire coast of England took hundreds of thousands of years to form. On the other hand, faults may propagate at the velocity of sound in rock (several kilometers per second) and movements on them may be measurable in meters per second. Huge flat-lying overfolds (recumbent folds or nappe folds) form in metamorphic rocks with stress differences between minimum and maximum compression of less than 100 bars (10 MPa) within a few million years. In contrast, brittle failure causing fault motion may require more than 1 kbar (>100 MPa) differential stress and occur in seconds.

The shapes that folds may take and the orientations of those folds give rise to a wide range of geometric possibilities that we may meet. In this chapter, we shall deal with folds at a rather simple level although the description of folds in the diagrams is somewhat more encompassing than we need and is given for your future reference. Even then, the simplification is made that the folds are regular geometrical forms; in particular, that they preserve the same degree of folding along their hinge-line length. Initially, we will consider only angular (chevron) folds since these have planar limbs (Figure 10.1(a)). Folds with curving hinges and curved limbs are considered later but a simple classification based on interlimb angle and axial plane dip is given in Figure 10.1(b and c). Initially also, we consider only folds that have a horizontal plunge, i.e., the fold axis is not tilted. Initially also, we restrict ourselves to upright folds, ones in which the axial plane is vertical. Thus, after reviewing the diagrams, we will first tackle the simplest fold map construction where the folds are

1. angular (=flat limbs, sharp hinge),
2. upright (=axial plane of symmetry vertical), and
3. nonplunging (=axis or hinge line is horizontal).

These diagrams will be reviewed but here we address the question that may have arisen earlier in your minds. Planar, nonfolded strata crop out along contorted patterns, dictated by topography. How then will we determine from map view, if the sinuous outcrop patterns are due to folding and topography or solely due to topographic effects?

This problem is not as great as it may seem. The following steps are useful.

1. Can a bend of strata be recognized anywhere?
   a. Consider the case where the topography is negligible, as in most regional maps such as scales of 1:100,000, 1:200,000, etc. In these cases, any substantial bends in a stratum will indicate the hinges of plunging folds. Where fold hinges are horizontal or plunge very gently, map view may not reveal the closure or curvature of the fold hinge. However, in that case, sequences of strata (e.g., A–B–C) will repeat in mirror image order (i.e., C–B–A) across the map.

   b. Where topography is a consideration, e.g., at scales of 1:50,000 or 1:10,000, reading the sense with which beds "climb" or "descend" slopes will reveal the presence of a fold. For example, even a fold with a horizontal axis will expose the closure (bend) of the hinge region where the fold axis meets a topographic slope. Inspect the trend of the strata; this must correspond broadly to their strike and broadly to the fold axis trend. Where the fold meets a slope at a high angle, the same bed may change elevation as it crosses the slope to form a downslope bend (synform) or an upslope bend (antiform). This is a robust map-reading approach that simply detects the presence of a fold by inspection. It rarely reveals fundamental information concerning the shape or orientation of the fold since the map view contortions of a folded bed are dependent of the topography. However, we shall see that simple constructions reveal quantitative information on the fold's shape and orientation. Eventually, the geologist learns to perform similar qualitative interpretations mentally.

2. Is the stratigraphic sequence repeated in two nearly opposite directions? A fold necessarily bends the strata so that the two flanks (limbs) tilt the strata in nearly opposite directions. When exposed by erosion, the sequence of the beds crops out in mirror image on either side of the fold's axis. For example, the beds get younger away from the core of an anticline; the beds get older away from the core of a syncline. Consider a sequence of beds, **A** is older than **B** which is older than **C**. Now, inspect the map; does the sequence **A→B→C** occur in one direction (say, younging eastwards) in one part of the map and does it then occur younging eastwards **C←B←A**, in another part of the map? This would indicate that the strata dipped in opposite directions due to folding.

3. *Are dip-and-strike symbols present?* On some maps, dip-and-strike symbols will provide the obvious clue to the presence of folds. An antiform separates obviously opposed dip directions and face-to-face dip directions occur on either flank of a synform. A further clue to the presence of folds may be provided by antiparallel, or nearly opposed, younging arrows, **Y** [younger beds at the foot of the letter symbol], mainly on maps of very complexly deformed rock. However, some caution may be required in the use of structural symbols, depending on the scale of the map and the size of the folds. For maps that cover very large areas (1:50,000 and 1:100,000), the symbol must be very carefully placed to avoid misleading the map reader since actual outcrops are smaller than the symbol. It should also be remembered that the symbol may cover a huge ground area, of several hundred square meters. A common convention is that the tip of the symbol locates the site of the observation. This does not help very much in many cases; I have

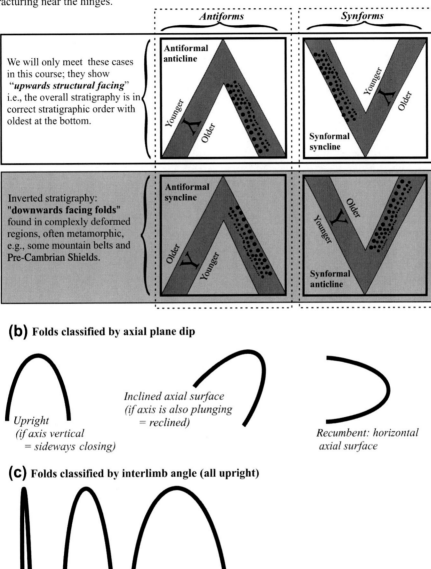

(a) Folds classified by stratigraphy and form

Idealized fold profiles of folds are shown; they are cross-sections perpendicular to the fold hinge. For simplicity, chevron ("angular") folds, with flat limbs are shown; in nature all folds show at least some curvature in the hinge regions. Field examples of natural chevron folds also show some fracturing near the hinges.

Antiforms *Synforms*

We will only meet these cases in this course; they show *"upwards structural facing"* i.e., the overall stratigraphy is in correct stratigraphic order with oldest at the bottom.

Antiformal anticline

Younger Older

Younger Older

Synformal syncline

Inverted stratigraphy: **"downwards facing folds"** found in complexly deformed regions, often metamorphic, e.g., some mountain belts and Pre-Cambrian Shields.

Antiformal syncline

Older Younger

Older Younger

Synformal anticline

(b) Folds classified by axial plane dip

Upright
(if axis vertical = sideways closing)

Inclined axial surface (if axis is also plunging = reclined)

Recumbent: horizontal axial surface

(c) Folds classified by interlimb angle (all upright)

Isoclinal *Tight* *Close* *Open*

Note: various combinations of axial plane dip, axial-plunge and tightness give rise to logical adjectival desscriptions, e.g., reclined open fold, axial surface dipping 30^0 to east, plunging 20^0 to south. (can you sketch or make a sketch map of such a fold?)

FIGURE 10.1 Classification of simple fold forms. Most natural folds have much more complex curving shapes (a) (b) and (C) identified in drawing.

seen regional maps (1:100,000 and larger) on which the area covered by the symbol is so large that it obscures the area in which the fold crops out. This is generally the reason why regional maps omit all small-scale structure symbols of any kind. Such large-scale maps also present a further difficulty; topographic relief is usually so small in comparison to the map scale that we cannot use topography–geology relations to deduce reliable geological relations such as dips and sequence. Such large-scale maps tend to become diagrams that only show geographical distributions of rocks. Three-dimensional interpretation requires some further skills.

4. *Construct stratum contours!* As with faults, the only reliable way to determine the presence, nature, and effects of folds is to construct stratum contours (Figure 10.2). This technique works where we have small scales and topographic contours. Structure contours will trace out the form of a bed as if it cropped out on a hypothetical horizontal plane (or mine plan). Clearly, the same stratum contour (e.g., sandstone–limestone, 350 m) will appear twice for each fold, once on each fold limb. With good "book keeping", by labeling each stratum contour carefully and working in erasable pencil, we may interpret such maps uniquely. A common elementary confusion is that when the intersections of topography and a bedding plane are determined, upon first sight there may appear to be ambiguity in the way they are connected. The correct of possibilities will be parallel to the respective flanks of the fold. The alternative set of misconstructed stratum contours will cut across the symmetry of the fold and be inconsistent with topography. Recall earlier that we learned that a geological boundary appears only at the topographic surface if is justified by the intersection of a topographic contour and stratum contour of the same elevation.

5. Where folds plunge (Figure 10.3), the stratum contours from opposing flanks will meet at the hinge (Figure 10.3(b and c)). The separation of the stratum contours along the axial trace permits the plunge of the fold to be determined. This is equivalent to determining the dip of the hinge of the fold along the axial trace (Figure 10.4).

Some natural structures flag the geometry of the major fold. For example, cleavage or schistosity is commonly parallel to the axial planes of fold (Figure 10.4(a)). In a similar fashion, minor folds on the flanks of major folds will define the plunge of the major fold (Figure 10.4(b)). Note that the asymmetry of the minor folds changes from one major flank to the other; commonly, these are termed "S" and "Z" asymmetry folds. However, their exact appearance on the map also depends on the slope of the topographic surface on which they appear; this will become apparent in later exercises (e.g., Figure 10.14).

Natural folds are not as simple or "perfect" as the examples in the preceding diagrams. An example of a natural fold is mapped in Figure 10.5. Note that the main synform is accompanied by two minor folds on its southern flank. However, these folds do not persist for a long distance along their axial traces. Such folds are termed "disharmonic" as minor axial traces eventually disappear, merge, or are faulted out.

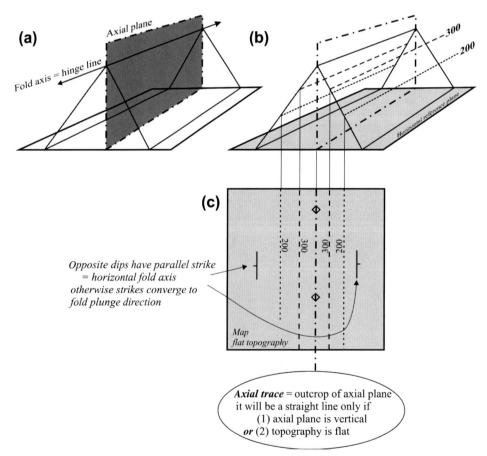

FIGURE 10.2 Isolation of folds (in this case an antiform) using structural contours. (a) Simple antiform (b) Same with stratum contours (c) Map view of structure contours.

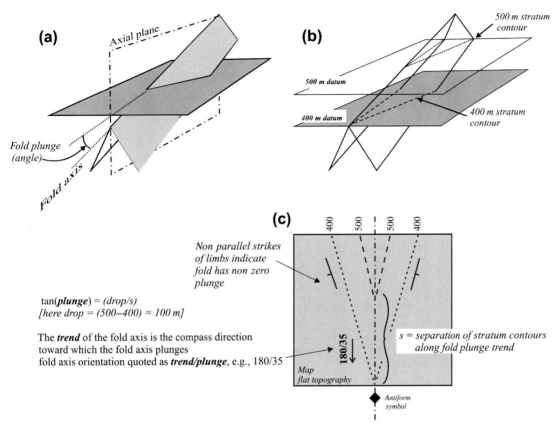

FIGURE 10.3 (a) A simple plunging angular antiform (b) structure contours (c) map view.

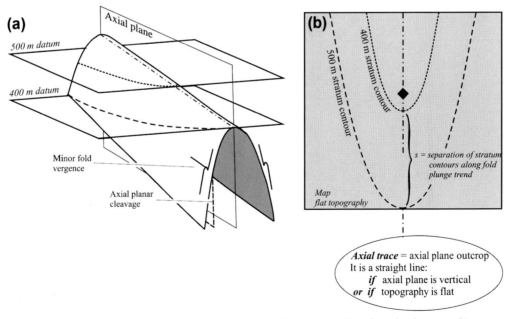

FIGURE 10.4 A rounded antiform; note the curving structural contours. (a) three-dimensional structure (b) map.

Some simple exercises with folds illustrate these concepts. First, in Figure 10.6, we meet upright folds without plunge; structure contours will be parallel and straight. Determine the stratigraphic order by examining topographic slopes (shale and sandstone are illustrated). Determine the dip angles at the locations marked with dip-and-strike symbols. Using labeled stratum contours, determine the presence of any folds. Mark the axial traces of

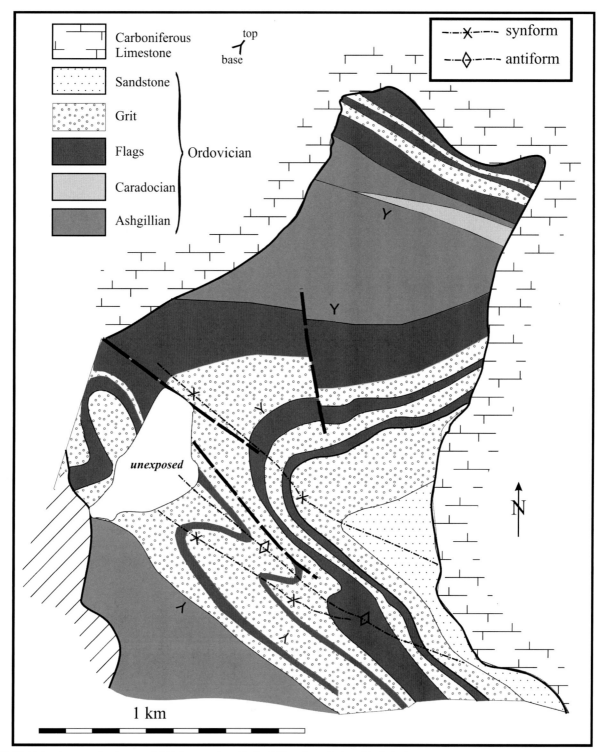

FIGURE 10.5 An example of a natural fold, from northern England. Note the variation of limb thicknesses and the fact that the minor folds are disharmonic, i.e., they vary in amplitude along their axial planes, causing them to disappear or merge.

the folds following the examples given. Draw a true-scale cross-section LR with L on the left, extrapolate the geology above and below the topography. The cross-section is perpendicular to the fold axes and thus it will show a true plunge profile.

Figure 10.7 shows an example of a plunging fold with flat hinges that give straight structure contours. However, the fold plunges so that the structure contours are not parallel, as illustrated. Complete the structure contours and extend them to the edges of the map. Determine the dips

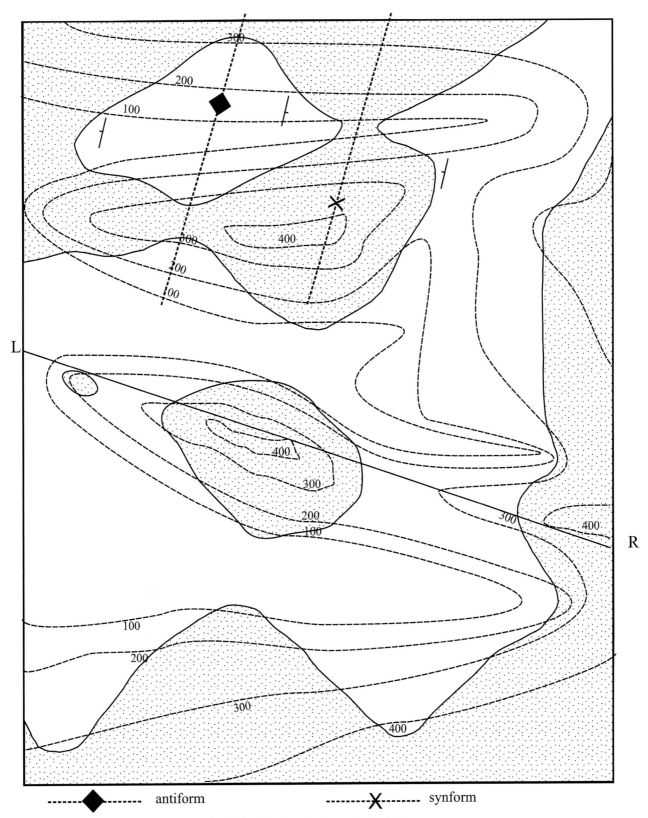

FIGURE 10.6 Example of nonplunging folds; see text.

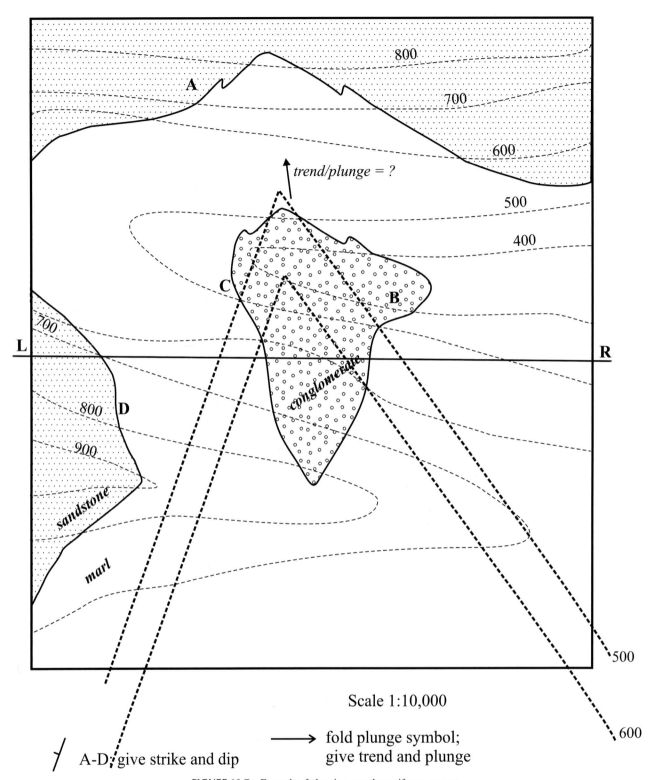

Scale 1:10,000

fold plunge symbol;
give trend and plunge

A–D, give strike and dip

FIGURE 10.7   Example of plunging, angular antiform; see text.

and strikes at locations A–D. From the spacing of the structure contours along the hinge line (=axial trace) determine the plunge angle of the major fold. Draw a cross-section from L (Left) to R, extending the geology below and above

the topographic surface. The hinge of the major fold crops out on the hillside to the right of A; its shape gives a clue to the curvature of the hinge in cross-section, i.e., it should not be sharply angular. Note the presence of minor folds near

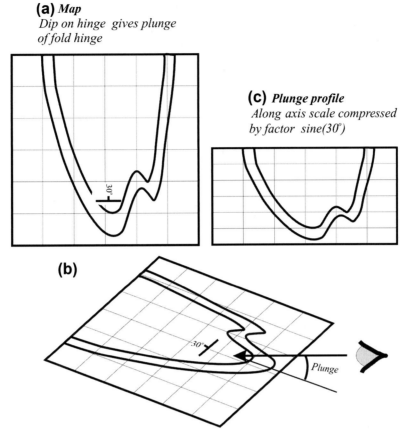

(a) *Map*
Dip on hinge gives plunge of fold hinge

(c) *Plunge profile*
Along axis scale compressed by factor sine(30°)

(b)

Plunge

FIGURE 10.8 Plunging folds are difficult to accurately represent in cross-section. The map (a) may be viewed down plunge (b) to reveal the true plunge profile. This may be constructed by redrafting the map on a compressed grid.

A, B, and C and their change in symmetry across the hinge of the major fold. Can you represent these in the cross-section? Note that the cross-section does not reveal a true plunge profile. A special construction as follows is required to reveal the profile.

Figure 10.8 shows how we may reveal the true appearance of a fold's plunge profile. (a) is a map of the folds with a 30° plunge (shown by the dip at the hinge). If one tilts the map and views the fold obliquely (b) at a 30° angle the true shape of the fold is revealed. This may be drawn by gridding the map (a) and redrawing the fold with a compression factor of sine (30°) as in (c). (c) is the true plunge profile. Now return to Figure 10.5 and produce a plunge profile of those folds.

Figure 10.9 introduces greater complexity. Major folds do not plunge but they are partly concealed by an unconformable cover. Cleavage symbols show the orientation of a vertical axial plane cleavage. First, mark the unconformities (trace them out). Determine the dips of strata at a–d and show them on the map. Mark the unconformity. Construct structure contours for the folded beds below the unconformity (there is only one bedding plane that can be used). Locate the axial traces of the folds and mark them on the map, beneath the unconformity. They are located between the central structure contours for the folded bed.

Draw a cross-section LR with L on the left, extend the geology above and below the topography. Make sure you calculate the apparent dips of strata correctly in the section using strike lines as per exercise (Figures 7.1 and 7.2). Can you draw the subcrop of the bedding plane beneath the unconformity?

Figure 10.10 provides a light relief from the previous exercises. Complete the map using the information shown in the EW sections at the top and bottom of the map. Note strike-and-dip symbols provide clues on the map. Note that the fold is not angular but gently curved.

Figure 10.11 requires you to complete the geology north of the fault but first, determine dips and strikes at A through E. Construct structure contours south of the fault, adjust them to determine the elevation of structure contours north of the fault (for simplicity, the throw is 100 m). Construct and label the structure contours north of the fault and complete the geology north of the fault. Determine the borehole logs at W and Y (i.e., the vertical sequence of strata at those locations.) Draw a cross-section left to right, LR, carefully calculating the apparent dips of strata using the method of Figure 7.2.

Figure 10.12 reveals upright folds with a fault history. Determine structure contours for the fault, determine its dip and construct structure contours for the stratum north of the

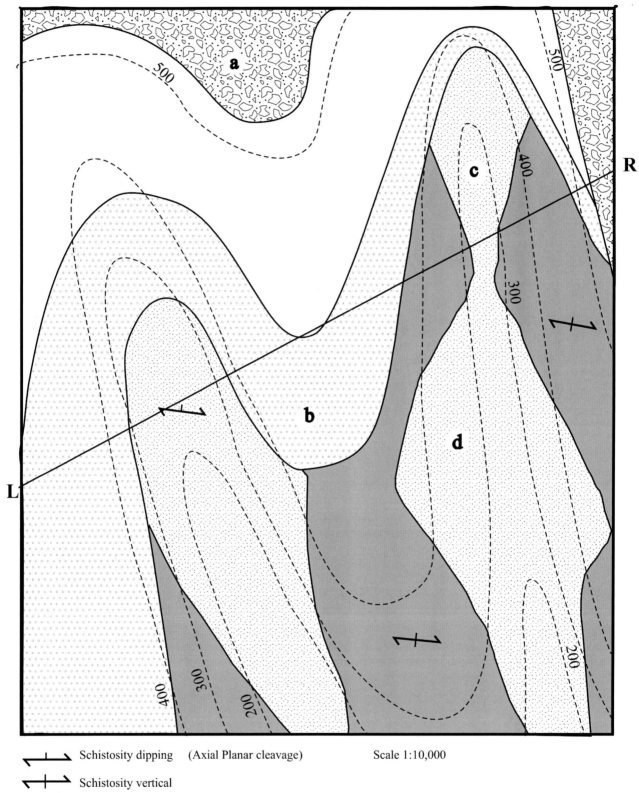

Schistosity dipping   (Axial Planar cleavage)   Scale 1:10,000

Schistosity vertical

FIGURE 10.9 Horizontally plunging (i.e., nonplunging) folds, partly concealed by an unconformity. Locate the axial traces of all folds and complete cross-section.

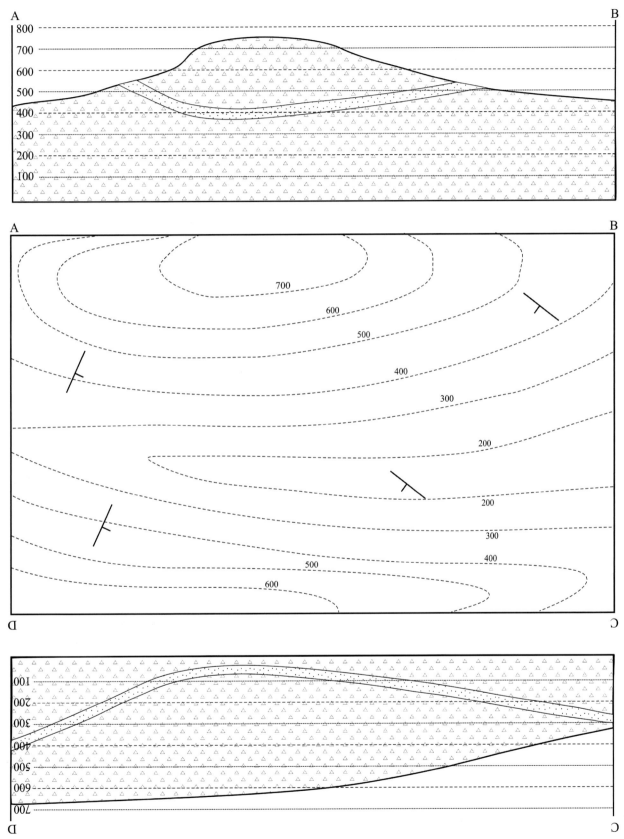

FIGURE 10.10 Complete the map of the plunging, curved fold, from the cross-sections.

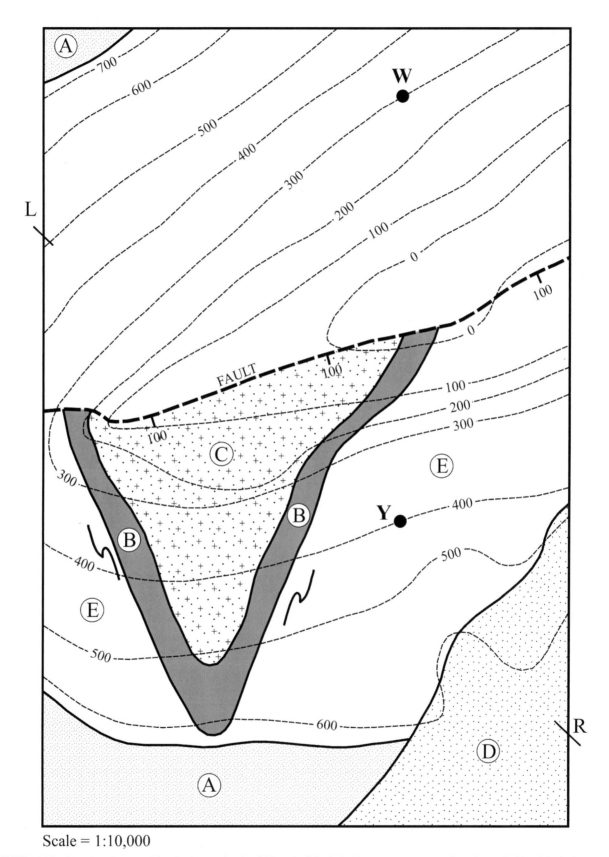

Scale = 1:10,000

FIGURE 10.11   Complete the map of the simple, nonplunging fold north of the fault. Use structure contours from the south side of the fault to map in the geology north of the fault.

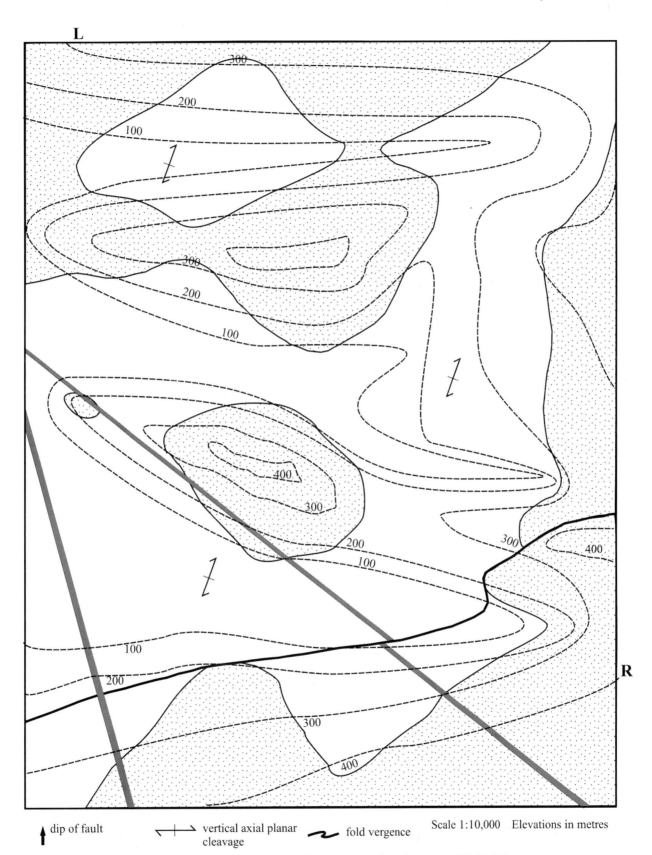

FIGURE 10.12   Locate the fold axial traces and determine the throw on the fault using carefully labeled structure contours.

fault. Construct structure contours south of the fault and determine the throw and nature of the fault. Draw a section left to right, LR. Explain the nature of displacement of the dikes. Mark the axial traces of the folds; they are straight lines since the folds are upright and parallel to the axial planar cleavage.

Figure 10.13 shows a series of simply folded beds beneath an unconformable cover. Trace out the unconformity, determine the dip and strike of the unconformity and overlying strata.

Construct structure contours for one horizon of the folded strata. Locate the axial trace of the fold(s) and indicate the dip and strike of the flanks. Do the folds plunge? Construct the subcrop of the folded sandstone horizon as per exercise (Figures 8.4 and 8.5). Construct a cross-section LR with L on the left. How does the angle of folding differ in this section from the true hinge angle?

The vergence (asymmetry) of minor folds is shown at the western edge of the area. Can you complete a suitable symbol for vergence at a–d?

Subsequent exercises will show minor folds and axial plane cleavage associated with the major folds on the maps. Commonly, minor folds are referred to as having a "Z" or "S" geometry on the flanks of a major fold (and an "M" shape on the hinge). An example is shown in Figure 10.14(b), where the minor folds flank an antiform. However, the view of the minor folds depends very much on the surface orientation; in (a), the folds both yield an S and a Z geometry but the interpretation is unique, an antiform lies to the south. For this reason, it is better to note the interpretation of the asymmetry as in (b), using a solid circle for the N side of the antiform and an open circle for the S side of the antiform. The map of the road section in (c) shows how the ambiguity may be resolved.

Other exercises will show axial plane cleavage; this is an important minor structure used in mapping (Figure 10.15). Also known as Pumpelly's rule, it basically states that the cleavage is, as a general rule, closer in orientation to the axial plane than the flanks of the fold. Thus, given outcrops P and Q, the only solution is that (at least one) antiform lies between the outcrops. The mechanical basis for this is that even at low metamorphic grades, the compression causing folding simultaneously causes metamorphic reactions that align minerals in the most compliant orientation (i.e., micas parallel to axial plane). (b) shows the interpretation of (a). (c) shows the interpretation of cleavage bedding angles on the left hand flank of an antiform and an overturned antiform.

Figure 10.16 introduces a fold with overturned limbs. The folds do not plunge so that the structure contours appear as approximately straight lines. (Some symbols for bedding indicate the strike direction.) What is the trend of the fold indicated? Draw structure contours on the quartzite–slate boundary, starting at the left hand edge of the page. Mark dip-and-strike symbols as you go and indicate the dip angles. You will first define an antiform trace that will curve in a

similar fashion to one shown on the left hand side of the map. However, it construction may be postponed at this stage. Complete the structure contours to define a synform between the two antiforms; note that the vergence symbols agree with the shapes of minor folds. Draw a section LR (left to right). Now draw in the axial traces of the folds. However, this is not a trivial procedure. Since the fold axial planes dip to the left, their axial traces have a curvature that must be in sympathy with topography. You can achieve this by drawing structure contours for the axial planes and construct the axial traces from their intersection with topography. Finally, note the vergence of the minor folds indicated for the fold in the slate near its boundary with the quartzite. In a similar fashion, draw the vergence of minor folds at the four locations marked with a heavy circle on the eastern antiform. Also determine the mean dip and strike at those locations and indicate it on the map. (Note the use of an inverted bed symbol.)

Figure 10.17 introduces overturned folds with cleavage and a fault. Locate axial traces of folds and indicate the dip of strata at various locations. Note the axial planar schistosity is a clue to the dip of the axial planes. Determine the dip and strike of the fault and its sense and amount of throw. What kind of fault is it? Draw a cross-section LR from left to right. Describe the structural history. On the cross-section, indicate the vergence (asymmetry) of minor folds at six locations. Can this be shown on the map?

Figure 10.18 shows a plunging, reclined (reclined = axial plane dips) fold of a sandstone bed. Construct the structure contours for one horizon on each flank of the fold. Where these meet at the same elevation, they will locate the hinge of the fold. Mark the dip and strike of the flanks on the map and the trend and plunge of the hinge. Draw a cross-section of your choice to reveal the nature of the fold. Construct the sinuous course of the axial trace across the map (since it dips it must be influenced by topography as in Figure 10.16). What is the true interlimb angle in the field and what is the apparent interlimb angle in your section? Using the symbol for the vergence of minor folds indicate the expected vergence on the map at locations a–d taking account of topography and major fold vergence. What is the trend and plunge of the major fold? Show this on the map with an arrow.

Figure 10.19 presents overturned but not plunging folds of a single horizon. Some field symbols show orientations of schistosity, bedding, and overturned bedding. Determine the axial traces of the folds as accurately as possible and draw a section LR from left to right. Using the symbol for the vergence of minor folds, indicate the expected vergence on the map at locations a–d taking account of topography and major fold vergence.

Figure 10.20 shows the structure of a complicated area. Note the minor fold symmetry and the orientation of beds. Describe the structural history as fully as possible using constructions to show your understanding. Indicate the dip angles at the locations of the dip-and-strike symbols.

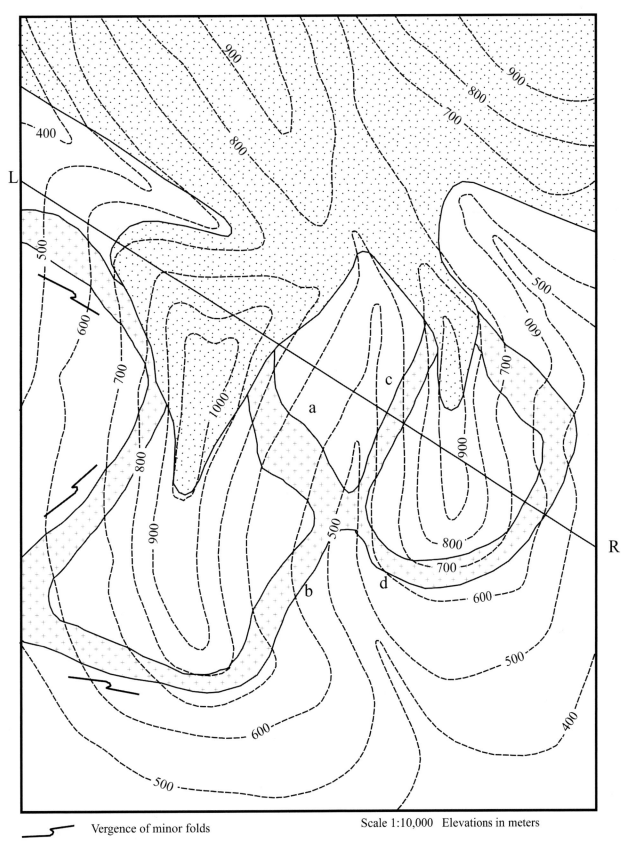

Vergence of minor folds   Scale 1:10,000   Elevations in meters

FIGURE 10.13   Construct structure contours for the folds, complete the subcrop of the folded sandstone beneath the unconformity. Draw a cross-section LR from left to right.

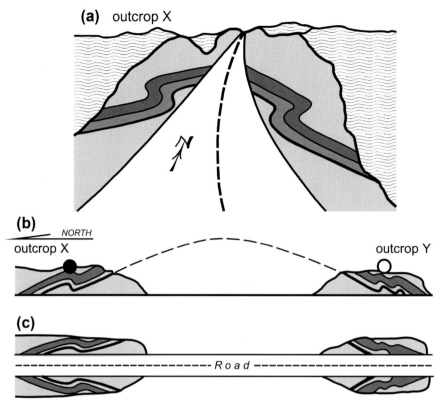

FIGURE 10.14 Subsequent maps will show minor folds on the flanks of the major folds. The asymmetry of the minor folds changes across the axial trace of the major fold but care must be taken to note the orientation of the surface on which the minor fold crops out. Thus, merely noting "S" or "Z" symmetry will not usually suffice. See text.

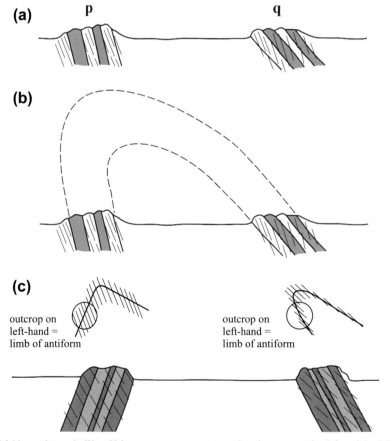

FIGURE 10.15 Overturned folds verging to the West. Using structure contours, complete the cross-section left to right (LR). Then, interpolate structure contours for the axial planes (their dip is intermediate between the limbs) and map in the axial traces. Add more minor fold vergences and indicate the orientation of axial plane cleavage.

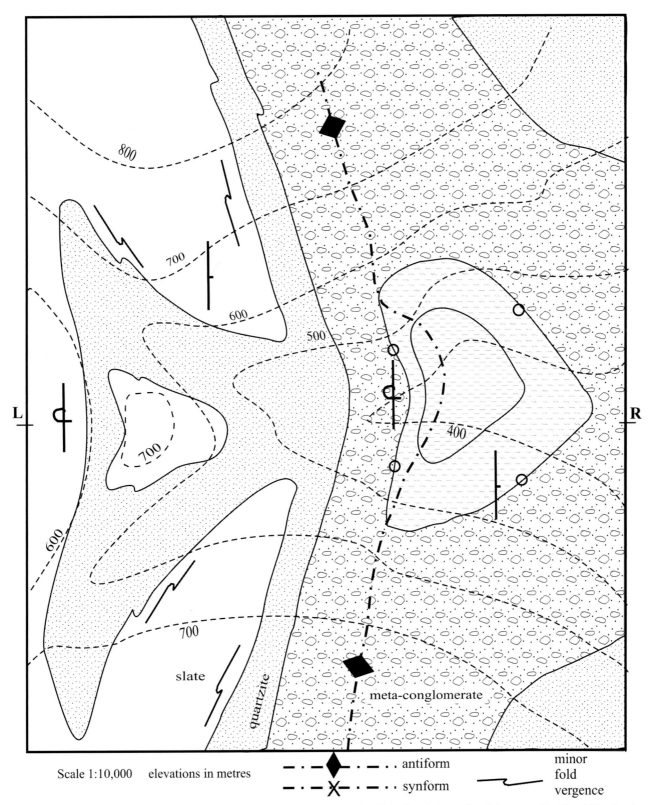

Scale 1:10,000   elevations in metres

— · — ◆ — · — · · antiform

— · — ✕ — · — · · synform

minor
fold
vergence

FIGURE 10.16 Some subsequent maps will show cleavage, axial planar to the major folds. The relative angles of cleavage and bedding (a) bracket the locations of major folds because cleavage is generally closer to the orientation of the axial plane (b).

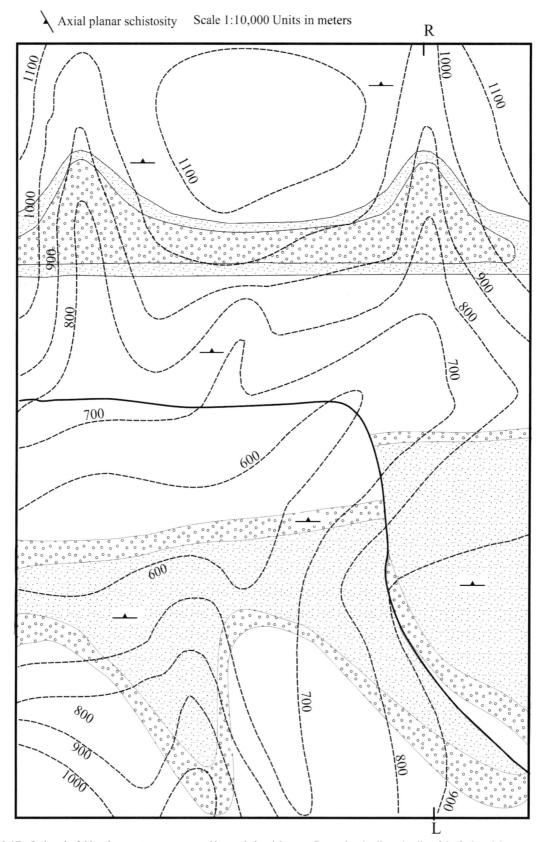

Axial planar schistosity   Scale 1:10,000 Units in meters

FIGURE 10.17   Isolate the folds using structure contours and locate their axial traces. Determine the dip and strike of the fault and the nature of it motion from the structure contours of the beds. Draw a cross-section left to right (LR). Add more indications of axial plane schistosity and estimate their dips.

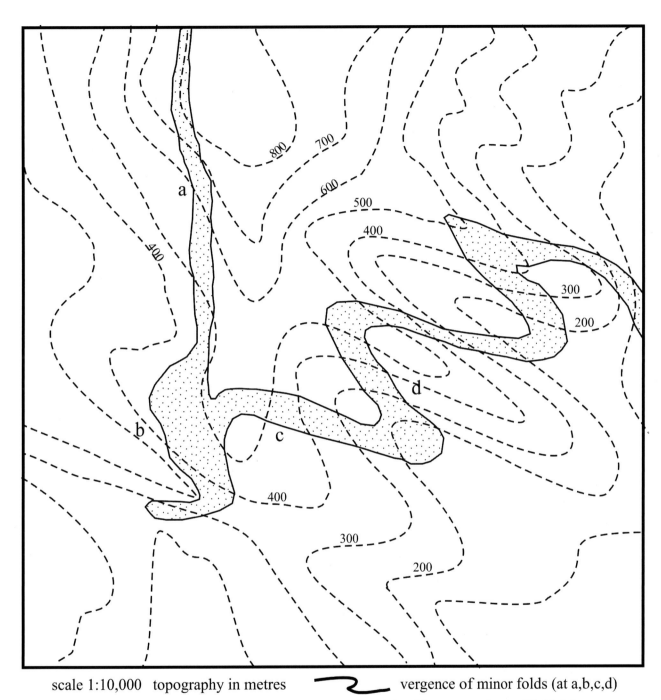

scale 1:10,000   topography in metres   vergence of minor folds (at a,b,c,d)

FIGURE 10.18 Plunging, overturned fold. To understand this fold and construct it axial plane, it is necessary to draw carefully labeled structure contours on one side of the bed. Indicate the dips of the flanks and the plunge of the fold. Mark in the axial trace bearing in mind its dip is midway between that of the flanks.

Figure 10.21 is a relatively simple review question. Determine the nature and motion of the fault, the location of fold(s), and the geological history. Locate the axial trace(s) of folds since the folds are upright, the traces are straight. A more complex question is to fix the vergence of minor folds, using the given symbol, at locations a–d. Be careful to take account of the vergence of the major fold and of the topography.

EXCERPTS SIMPLIFIED FROM PUBLISHED GEOLOGICAL SURVEY MAPS

In all of the maps, the area is sufficiently large that folded strata are recognizable without topographic interpretation. Thus, topography is not shown and the use of structure contours is not required. However, in some maps, such as the Bristol sheet, very gently dipping strata have tortuous outcrops due

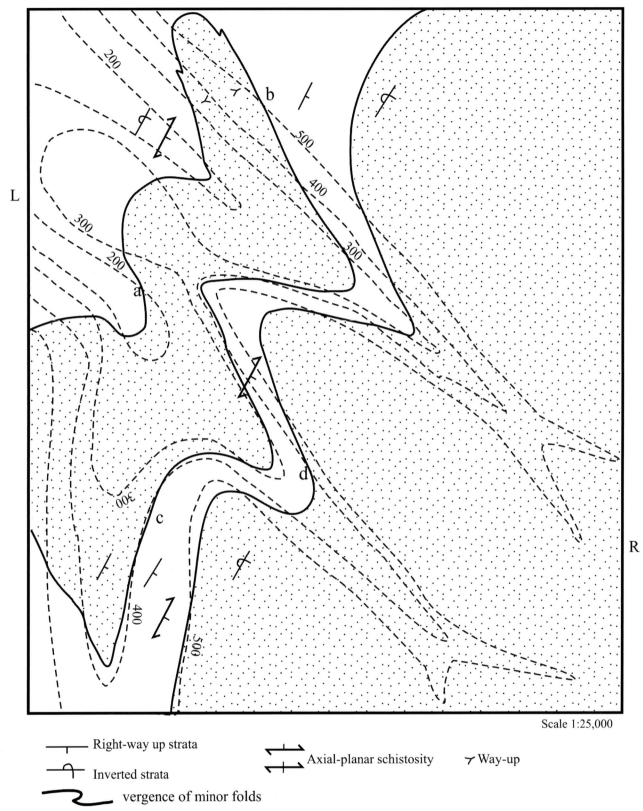

Scale 1:25,000

Right-way up strata

Inverted strata

Axial-planar schistosity   Way-up

vergence of minor folds

FIGURE 10.19 Overturned fold. Draw structure contours carefully for one bedding plane. Determine the dips of the beds and the plunge of the fold. Indicate the axial trace.

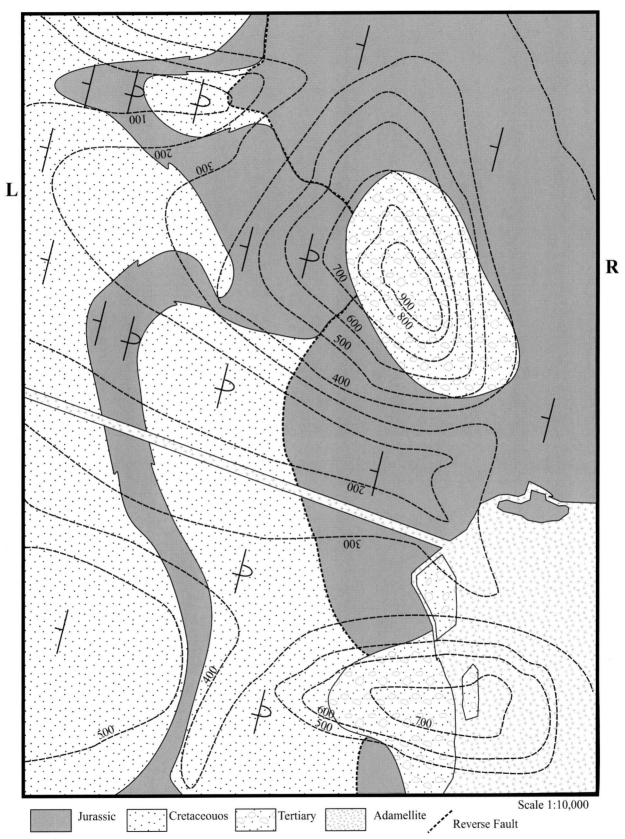

FIGURE 10.20 Overturned fold. Determine the geological history of the map and produce the cross-section.

Jurassic   Cretaceouos   Tertiary   Adamellite   Reverse Fault

Scale 1:10,000

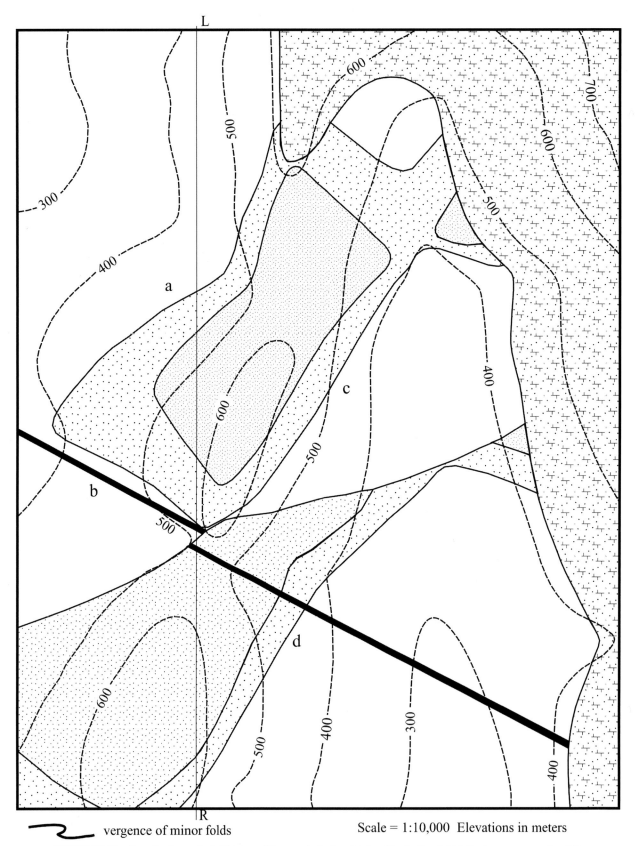

vergence of minor folds   Scale = 1:10,000   Elevations in meters

FIGURE 10.21   Describe the geological history of the area, including details of the fault and fold. Produce a cross-section LR.

to topographic effects. Their dendritic pattern indicates their superficial and horizontal nature. This area is affected by pre-Triassic folding due to the Variscan (=Hercynian) orogeny.

Bristol District SW, UK

1. The map has a scale that dwarfs the topographic relief, however, the very flat-lying strata do show some correlation with topography; namely, they follow dendritic drainage patterns (Figure 10.22).
2. On the map, indicate approximately, with pencil, the axial traces of any major folds. Use the appropriate symbol to distinguish antiforms and synforms. Use appropriate terms to describe the tightness, style, and orientations of the folds (e.g., closed, reclined, S-plunging, verging west; verging = axial plane leaning toward).
3. Draw an N–S cross-section along the section line drawn near the left margin of the map. Keep south on the left-hand side. Indicate the orientations of strata to some depth so that your interpretation is clear. Keep the section true scale, only then may you use dip angles from the map directly in the section. However, remember that if strata do not dip in a direction parallel to the section, they will show reduced apparent dips (recall Figures 7.1 and 7.2).
4. Establish a stratigraphic column that includes all geological events of significance. Identify the relative age and type of the principal stratigraphic discontinuities within the column. Also, indicate within the stratigraphic column where folds and different kinds of fault may have occurred.
5. Identify how many systems of faulting are present, by type and by orientation. What are the principal senses of motion of each group you identify? What are their ages of the groups identified, relative to folds and stratigraphic breaks?

East Falkland Islands, South Atlantic

East and West Falkland are parts of the ancient Gondwana continent, which were located close to the south pole during the Carboniferous, witnessed by the formation of the Fitzroy tillite (Figure 10.23). A tillite is a lithified till. Subsequent folding and overthrusting occurred followed by shallow water deposits of the Lafonia Group. The geology is very similar to parts of South Africa to which the Falklands were adjacent in Gondwana land.

1. This area is even larger than the previous map so that topography plays no part in the distribution of lithological boundaries.
2. Draw a cross-section LR with L on the left showing the orientation of the strata. Note that some younging arrows point in the opposite direction to the dip of the beds, meaning that the beds (and fold limbs) are overturned.

North West Cyprus

This large area of the NW part of the island of Cyprus in the eastern Mediterranean exposes an ophiolite, of Cretaceous age (mantle sequence approximately 88 Ma) (Figure 10.24). The mantle sequence of gabbros in the south center of the map forms the high point but the distribution of lithologies is, in the first instance, not topographically controlled at this scale. Rather, the mantle sequence and surrounding dike complex form an elongate dome running ENE–WSW. The doming is due to N–S shortening and perhaps due to hydration and volume increase of the mantle rocks at depth. The Paleogene limestone sequence drapes over the ophiolite complex and represents the original ocean floor sedimentary cover. NE of Morphou, a different tectonic terrane occurs representing a young fold and thrust sequence. In this case, different dip-and-strike symbols are used for the primary layering of sedimentary and volcanic igneous rocks. North of the ophiolite, east of Morphou, and to the north coast, a fold and thrust belt is exposed. What is its sense of vergence?

Draw a cross-section, ignoring topography, from L (left) to R (right).

Pierre Greys Lakes, Alberta

This map of tight folding covers a large area so that lithology controls the fold outcrop pattern not topography (Figure 10.25). The area was folded and metamorphosed during the Laramide Orogeny, approximately 70–40 Ma ago. Note that the lithologies are distributed in a regular elongate "grain" since there is a single tight episode of folding. Draw a cross-section of this fold and thrust belt from L to R, with R being on the side toward which thrusting and overfolding occurred. To which direction do the folds verge (vergence = direction to which axial planes lean)? Which is the direction of overthrusting?

Pembroke East, S. Wales

Note that the symbol used for overturned strata is now antiquated (Figure 10.26). The overturned folds have low plunges and were formed in the Hercynian (=Variscan) orogeny, a pre-Permian event. Add axial traces to the principal folds. Note also the pronounced "grain" to the geology as in the previous figure. At this scale, topographic relief does not influence the outcrop pattern. Draw a cross-section LR with L on the left. Some faults are thrusts. Which faults are thrusts and what is the direction of their overthrusting? In which direction, do the folds verge? (Is that apparent from the cross-section?)

Gananoque Area, Ontario

This area has tightly and isoclinally folded gneiss, deformed in the Grenville orogeny at approximately 1100 Ma (Figure 10.27). These rocks are nonconformably overlain by

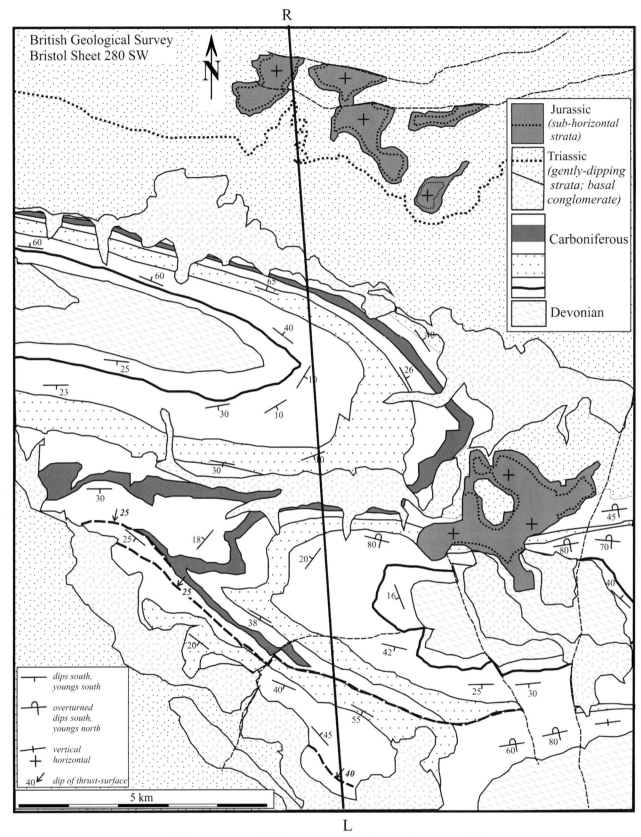

FIGURE 10.22 Major folds beneath an unconformity in the Bristol area of UK.

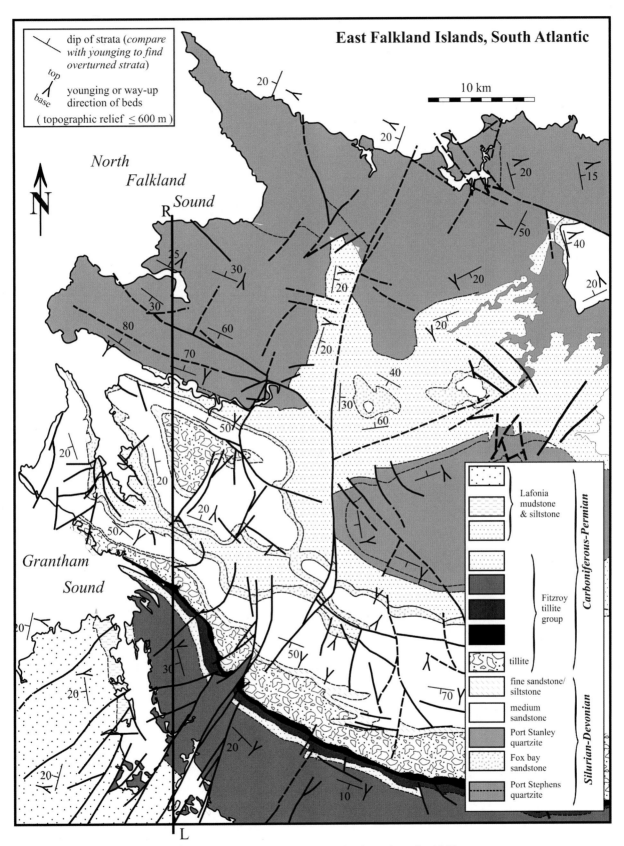

East Falkland Islands, South Atlantic

dip of strata (*compare with younging to find overturned strata*)

top
base
younging or way-up direction of beds

(topographic relief ≤ 600 m)

North Falkland Sound

Grantham Sound

10 km

Lafonia mudstone & siltstone

Fitzroy tillite group

tillite

fine sandstone/ siltstone

medium sandstone

Port Stanley quartzite

Fox bay sandstone

Port Stephens quartzite

Carboniferous-Permian

Silurian-Devonian

FIGURE 10.23 East Falklands geology showing major regional folds.

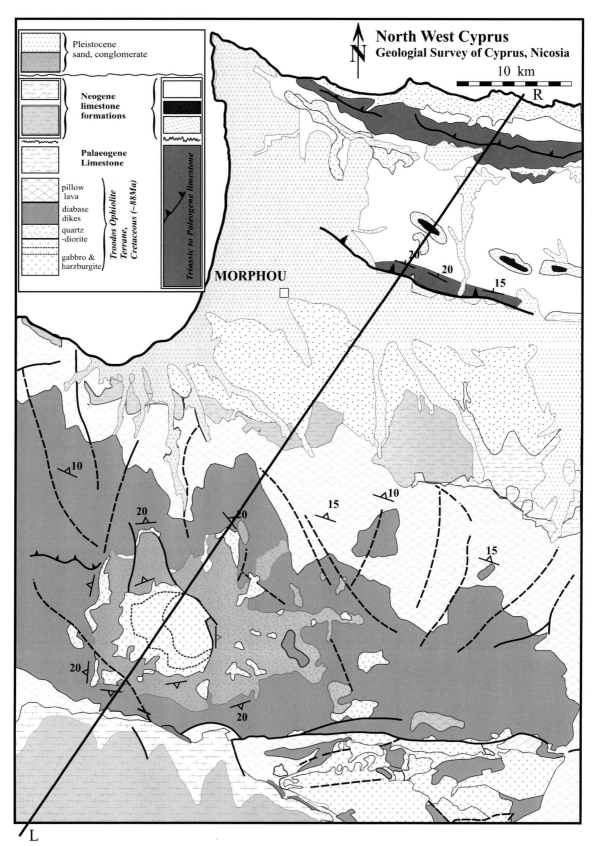

FIGURE 10.24 West Cyprus showing a major dome of an ophiolite in the south and a fold and thrust belt in the north.

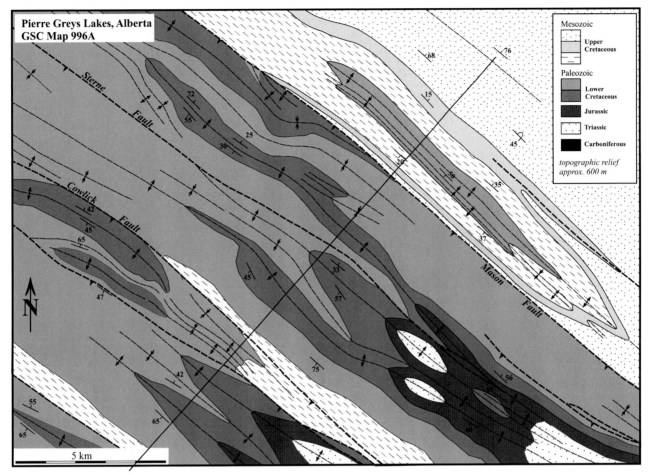

FIGURE 10.25 Pierre Greys Area, Canada showing a fold and thrust belt.

nondeformed Ordovician sandstone and dolomite. Trace out the unconformity. High grade of metamorphism and intense heterogeneous strain obscures the stratigraphic order of the Pre-Cambrian rocks. The orientation of the first schistosity (S_1) is given by the dip and strike symbols. This is axial planar to the first phase of folding (F_1). You will see that a second phase of folding (F_2) has deflected the earlier folds, after which gneissose granite was emplaced. Although the folding is tight or isoclinal, the "grain" to the geology is poorly developed because the phases of folding interfere with one another and are on a similar scale.

Label some fold axial planes F_1 and F_2 as appropriate. Tabulate a geological history for the area and draw a sketch cross-section L (left) to R.

Figure 10.28 shows an area of low topographic relief in England. Describe the fold and its orientation and construct a plunge profile as explained in Figure 10.8. What is the significance of the orientation of the cleavage–bedding intersection lineations? Slickensides are grooves or growth fibers related to bedding plane slip and are generated as bedding planes slide past one another during folding.

Figure 10.29 shows a map of the Loch Awe Syncline (LAS) and two adjacent minor folds. The Loch Awe Syncline is a major fold in SW Scotland affecting the Eocambrian Dalradian Series. Her, the Dalradian is unconformably overlain by Devonian lava. Draw a cross-section left to right (LR) and also produce a plunge profile of the structure. Is the Loch Awe Syncline overturned? Is its cleavage exactly axial planar? (See Figure 10.15.)

Figure 10.30 maps the contact between the Dalradian Supergroup and the older Moine Supergroup in Donegal, Ireland. The vergence of minor folds is shown by the black dot/open circle convention of Figure 10.14; larger minor folds are shown by the sinuous path of layering (foliation) which is termed an L–S fabric (L=lineation; S=schistosity). The contact between the Dalradian and Moine is a tectonic slide that is a shear zone with a normal sense of displacement.

Draw a plunge profile of the folding using the average mineral lineation as the plunge of the folds.

The folds in Figure 10.30 and some earlier diagrams show curvature of the fold hinges. Such folds are termed

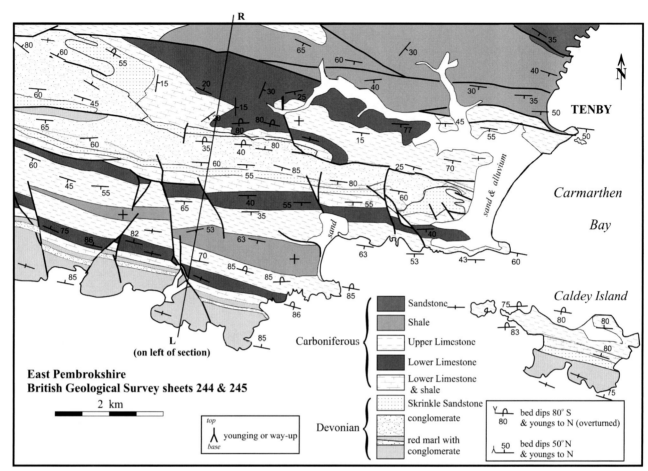

FIGURE 10.26 East Pembrokeshire, UK, showing folds and thrusts.

periclinal folds (or whaleback folds) and they produce variations in axial plunge and the orientation of cleavage–bedding intersection lineations. Figure 10.31(a) shows the initiation of such a fold due to heterogeneous flattening across the axial plane. A more extreme case is shown in (b) in which case the folds would be called sheath folds; younging directions are highly dispersed. An actual example of natural sheath folds is shown in (c) from Atikokan in N. Ontario. The section is approximately 10 km long, very well exposed, and abundant graded bedding and cleavage bedding relations reveal the structure as shown.

Figure 10.32 is a hypothetical map of folds with axial plane cleavage. Draw a cross-section LR showing the attitude of minor structures (folds and cleavage) and then produce a plunge profile.

Figure 10.33 is a famous cross-section of the SW Scottish Highlands mapped by Sir Edward Bailey (1934).

The section shows an isoclinally folded stratigraphy, but the isoclinals F_1 recumbent folds have been refolded by upright F_2 folds. First, trace in pencil the axial traces of the F_1 and F_2 folds across the section. Second, in the box below the cross-section make a sketch map of the geology in the section. Add symbols to show the orientation of bedding and first cleavage. The structural facing is the component of young projected onto the axial plane first cleavage; it yields sense of stratigraphic sequence from a single outcrop.

This area of negligible topographic relief occurs near Shebandowan, N. Ontario and was mapped by W. M. Schwerdtner (1984) (Figure 10.34). It reveals a pluton that has been refolded as revealed by the orientations of deflected L and S fabrics. Draw a plunge profile of the F_2 fold and mark on it the orientation of refolded S_1. Are the minor folds at the hinge of the pluton's major fold F_1 or F_2?

FIGURE 10.27 Gananoque East area, Canada, showing multiply folded and metamorphosed rocks of the Canadian Shield (Map after Hewitt, 1964).

FIGURE 10.28 Plunging anticline of Pre-Cambrian slates in Eastern England. Produce a plunge profile (see text).

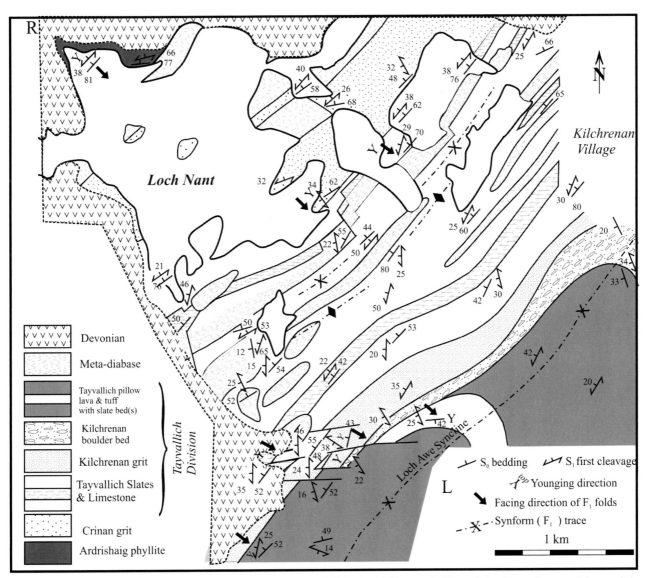

FIGURE 10.29 Overturned synform in the south (Loch Awe Syncline, SW Scotland.) Cleavage–bedding relationships and way up help reveal the structure (draw a section LR).

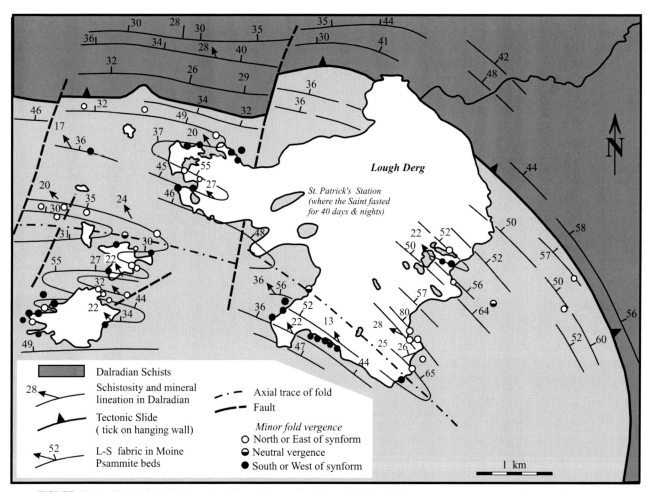

FIGURE 10.30   Tightly folded high-grade schists and psammites, Donegal. Minor fold asymmetry (vergence) reveals the major structure.

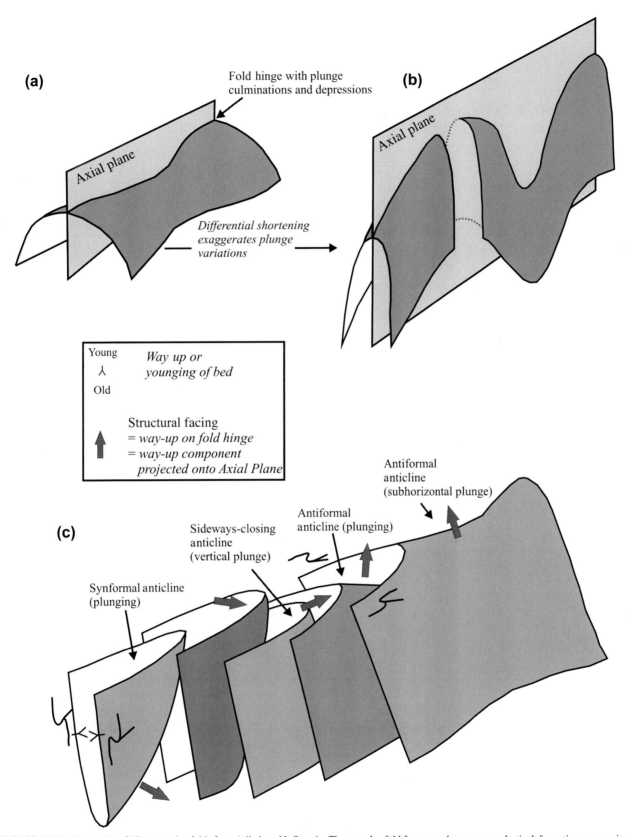

FIGURE 10.31 Examples of kilometer size folds from Atikokan, N. Ontario. The complex fold forms are due to severe plastic deformation progressing from the upper diagrams (a and b) to the final state in the lower diagram (c).

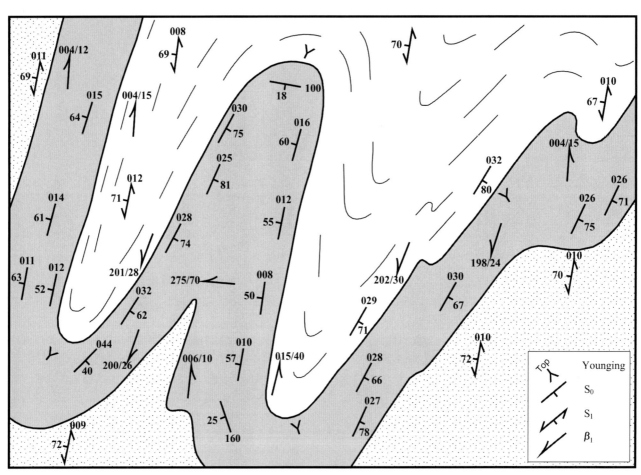

FIGURE 10.32   Hypothetical folding. Draw a plunge profile.

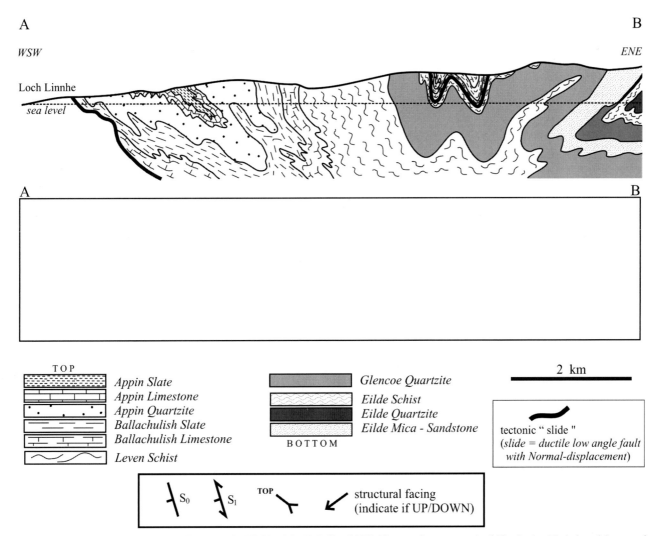

A   B

WSW   ENE

Loch Linnhe

sea level

A   B

TOP

Appin Slate
Appin Limestone
Appin Quartzite
Ballachulish Slate
Ballachulish Limestone
Leven Schist

Glencoe Quartzite
Eilde Schist
Eilde Quartzite
Eilde Mica - Sandstone

BOTTOM

2 km

tectonic " slide "
(slide = ductile low angle fault
with Normal-displacement)

S_0 S_1 TOP   structural facing
(indicate if UP/DOWN)

FIGURE 10.33   Cross-section of part of SW Scottish Highlands by E. Bailey (1934). The area shows two major fold episodes. Mark the axial traces of the F_1 and F_2 folds and complete a sketch map in the box below.

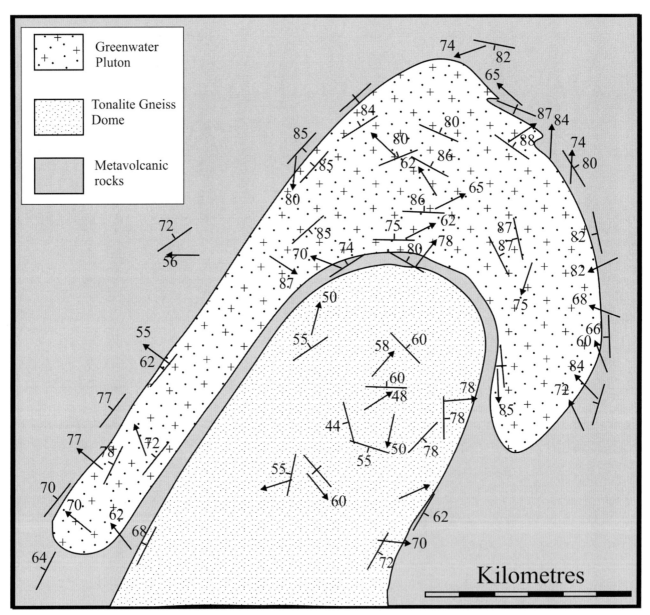

FIGURE 10.34   Map of NW Ontario by W. M. Schwerdtner showing two major phases of folding.

Cross-section **template**

Name (printed)...
Student # ....

Final Project Possible after Completion of Studying This Book

The project requires the construction of a hypothetical geological map with at least one cross-section, which clearly illustrates as much of the structure as possible. The final map must be neatly drafted and accompanied by several tracings that show the structure contours and construction of each structural feature and of each differently dipping sequence of strata. Show the want or duplication of one layer caused by the fault on a separate tracing. Marks are given for presentation and clarity as well as for accuracy. The topographic base map that you must use is attached, at a scale of 1:10,000, with elevations in meters. North is toward the top of the page.

The specifications for the geological structures that you must map are given in the following stratigraphic column. Commence by constructing structure contours for the oldest feature and proceed to add the younger elements in sequence. Do this on separate tracings. Finally, devise a topography, which must be geomorphologically reasonable, that reveals each of the elements of the mapped area in a fashion that permits the reader to determine their orientation and relative age. The map must also carry appropriately placed symbols to show the orientation of structures and, in the case of the fault, its throw also. Submit this table with your map and other materials.

Geological feature	Thickness or throw	Orientation (My name = ... My student # = ...)	
		Dip angle in degrees	Dip direction in degrees (show your values)
Jurassic shale	Top unknown	10°	50° + [3 × first nonzero digit of your student number] = ...
Cretaceous limestone	50 m		
Angular unconformity			
A dike	10 m	70°	35 × [last nonzero digit of your student number] = ...
A short sill apophysis from the dike	≤10 m		
Normal fault	100 m	60°	Your choice = ...
Cambrian slate	100 m	30°	100° + [10 × last nonzero digit of your student number] = ...
Nonconformity			
Archean gneiss	Base unknown		

Index

Note: Page Numbers followed by "f" indicate figures; "t", tables.

Printed in the United States
By Bookmasters